ENERGY POLICIES, POLITICS AND PRICES

THE ROLE OF TAXES IN ENERGY PRODUCTION AND CONSERVATION

ENERGY POLICIES, POLITICS AND PRICES

Additional books in this series can be found on Nova's website under the Series tab.

Additional E-books in this series can be found on Nova's website under the E-books tab.

ENERGY POLICIES, POLITICS AND PRICES

THE ROLE OF TAXES IN ENERGY PRODUCTION AND CONSERVATION

BERNADETTE M. HORTON
EDITOR

Nova Science Publishers, Inc.
New York

For permission to use material from this book please contact us:
Telephone 631-231-7269; Fax 631-231-8175
Web Site: http://www.novapublishers.com

LIBRARY OF CONGRESS CATALOGING-IN-PUBLICATION DATA

The role of taxes in energy production and conservation / editor, Bernadette M. Horton.
p. cm.
Includes index.
ISBN 978-1-61728-886-9 (hardcover)
1. Energy policy--United States. 2. Energy conservation--United States. 3. Tax incentives--United States. 4. Emissions trading--United States. I. Horton, Bernadette M.
HD9502.U52R635 2010
333.790973--dc22
2010026972

Published by Nova Science Publishers, Inc. ✢ New York

CONTENTS

PREFACE

Energy tax policy involves the use of one of the government's main fiscal instruments, taxes (both as an incentive and as a disincentive) to alter the allocation or configuration of energy resources and their use. In theory, energy taxes and subsidies, like tax policy instruments in general, are intended either to correct a problem or distortion in the energy markets or to achieve some economic (efficiency, equity, or even macroeconomic) objective. In practice, however, energy tax policy in the United States is made in a political setting, being determined by the views and interests of the key players in this setting: politicians, special interest groups, bureaucrats, academic scholars, and fiscal dictates. As a result, enacted tax policy embodies compromises between economic and political goals, which could either mitigate or compound existing distortions. This book explores the role of taxes in energy production and conservation.

Chapter 1- Energy tax policy involves the use of one of the government's main fiscal instruments, taxes (both as an incentive and as a disincentive) to alter the allocation or configuration of energy resources and their use. In theory, energy taxes and subsidies, like tax policy instruments in general, are intended either to correct a problem or distortion in the energy markets or to achieve some economic (efficiency, equity, or even macroeconomic) objective. In practice, however, energy tax policy in the United States is made in a political setting, being determined by the views and interests of the key players in this setting: politicians, special interest groups, bureaucrats, academic scholars, and fiscal dictates. As a result, enacted tax policy embodies compromises between economic and political goals, which could either mitigate or compound existing distortions.

The economic rationale for government intervention in energy markets is commonly based on the government's perceived ability to correct for market failures. Market failures, such as externalities, principal-agent problems, and informational asymmetries, result in an economically inefficient allocation of resources in which society does not maximize well-being. To correct for these market failures governments can utilize several policy options, including taxes and regulation, in an effort to achieve policy goals.

Current energy policy reflects efforts to achieve both current and past policy objectives. Recent legislative efforts have primarily focused on renewable energy production and conservation to address environmental concerns. In contrast, past efforts attempted to reduce reliance on foreign energy sources through increased domestic production of fossil fuels. Recently enacted legislation focusing on encouraging renewable energy production and conservation reduces reliance on imported, foreign oil, while also addressing environmental concerns by reducing the use of fossil fuels. Favorable tax preferences given to domestic fossil fuel energy sources also promote domestic energy production, reducing the demand for imported oil.

Chapter 2- The Senate Committee on Finance has scheduled a public hearing on April 23, 2009, titled "Technology Neutrality in Energy Tax: Issues and Options." Since 2004, the Congress has been very active in promulgating legislation related to energy production (including oil and gas and renewables) and conservation. Part II of this document,[1] prepared by the staff of the Joint Committee on Taxation, provides a description of present law tax expenditures for energy production and conservation. These tax provisions vary widely, and are summarized in the tables that follow. In addition to the energy specific tax expenditures, energy sector producers and manufacturers may also benefit from other general tax expenditures, such as the section 199 deduction for income attributable to domestic production activities, additional first year depreciation ("bonus depreciation"), the lower rates of tax on the first $10 million of corporate taxable income, and, in general, the deferral of tax on active business income derived by foreign subsidiaries.

The various tax benefits create incentives that have the potential to affect economic decisions and allocate economic resources from other uses to the tax-favored uses. Such tax preferences may produce an allocation of resources that is more efficient for society at large if they are properly designed to overcome negative effects (such as atmospheric pollution, for example) that would otherwise result from a purely market based outcome without any government intervention. The extensive variety of tax expenditures for energy production and conservation have been criticized for lacking well defined objectives, and for lacking coordination among provisions having similar objectives. Some argue that the simultaneous existence of tax preferences for the fossil fuel industry and for renewable energy production represents an incoherent government policy. Others have noted that the incentives for renewable energy and conservation are not themselves designed in a coordinated way to produce the most efficient or equitable subsidies for renewable energy and conservation.

Part III of this document provides the staff of the Joint Committee on Taxation's most recent estimates of tax expenditures in the energy sector, a brief discussion of the economic rationale for certain government intervention in energy markets through the tax code, and issues related to the proper design of such tax preferences.

Chapter 3- Justice Oliver Wendell Holmes wrote: "[T]he best test of truth is the power of the thought to get itself accepted in the competition of the market. . . ."

Today we consider a new thought in the marketplace of ideas about tax incentives. The thought is that, in the creation of energy tax incentives, the government might not pick and choose among different technologies.

The thought is: The government might just set a performance standard, regardless of the technology employed. We could encourage things like reduction in greenhouse gas emissions, improvement in efficiency, or increased energy content. And then we would leave the job of picking the best technology to the competition of the market.

Folks call this sort of incentive structure "technology neutral."

There are several reasons why a tech-neutral approach to energy tax incentives might make sense.

First, it might well provide more bang for our energy-tax buck. By tying receipt of these credits to a common standard, we may be able to set the level of incentives more efficiently.

Second — and this may come as a surprise to many in the room — sometimes government gets it wrong.

Consider the credit that Congress enacted in 1980 to stimulate oil shale, tar sands, and synthetic fuels from coal. The idea sounded good at the time. But many companies exploited the credit. Some sprayed coal with chemicals for no reason other than to line their pockets.

Third, a technology-neutral approach might mean that we have to change the tax code less often. Technology is bound to change faster than Congress can act. So there's appeal to instituting a series of incentives that we don't have to update all the time.

We've already taken some steps toward technology neutrality in recent legislation.

For example, as part of last year's farm bill, Congress enacted the first-ever credit for cellulosic biofuels. Now this might sound like credit that picks a specific technology. But in fact, we passed a technology-neutral credit for cellulosic biofuels

Chapter 4- As hockey and soccer players know, it's important to have goals. As we look at whether it makes sense to design technology-neutral energy tax incentives, we first need to consider what goals our energy tax policies seek to achieve. Some of the goals that have been mentioned for energy tax policy are a reduction in dependence on foreign oil, a reduction in the use of fossil fuels, and a reduction in carbon dioxide emissions. Depending on what goal or goals are selected, vastly different results emerge. For instance, if the only goal in the fuels arena is a reduction in dependence on foreign oil, then energy tax incentives that encourage more domestic drilling and oil production are appropriate. However, if the goal is solely a reduction in carbon dioxide emissions or a reduction in the use of fossil fuels, then those same energy tax incentives to encourage more domestic drilling and oil production are inappropriate.

Simplifying the energy-tax incentives by creating technology-neutral tax incentives is a noble ambition. However, getting consensus on what goals should be used in developing energy tax incentives can be a little like herding cats. Even if lawmakers agree on what goals should be used, which is a big "if" controversial issues arise. For example, whether nuclear energy should qualify for technology-neutral energy tax incentives would certainly be a controversial issue.

Also, the energy tax incentives that the Finance Committee has developed over the years have been extremely successful. For instance, the wind industry in the United States has made great advances with the help of the production tax credit, which I first authored in the early nineties.

Similarly, VEETC has helped the ethanol industry to reduce our dependence on foreign oil, improve our national security, and reduce carbon-dioxide emissions.

Chapter 5- Chairman Baucus, Senator Grassley, and Members of the Committee, thank you for the invitation to testify this morning on the issue of technology neutrality in the treatment of energy in the tax system. I make the following points in my testimony today

- Energy policy is shaped in important ways by the federal tax system. While taxes are one instrument of tax policy, subsidies in the form of accelerated depreciation, percentage depletion, production tax credits and investment tax credits are more commonly used instruments in the tax code.
- Technology neutrality can be defined in a variety of ways. It can be defined in terms of the effective tax rate on new investments in the sector, in terms of the levelized

cost of power from new investments or in terms of specific policy goals that motivate energy tax incentives.

- Efficiency is best achieved by setting taxes on energy sources that have negative externalities associated with their production or consumption. Similarly technological neutrality is most easily achieved through the use of taxes.
- A second-best technological neutrality can be achieved through the use of subsidies but it is more difficult to do so. In particular it is very difficult to level the playing field across different non-polluting energy sources through the use of subsidies.

Chapter 6- Good morning Mr. Chairman, Senators and distinguished guests. Thank you for the opportunity to offer my views on the pros and cons of technology-neutral energy and environmental incentives. I will confine my remarks to the area I know best, which is the transportation sector, addressing primarily incentives for energy efficient and low greenhouse gas (GHG) emitting vehicles.

For most energy and environmental policy goals, performance-based, technology-neutral incentives (and standards) are superior to those that target a specific technology. Performance- based incentives allow the widest scope for innovation, and permit market forces the greatest latitude to select and implement cost-effective solutions. Because of our limited ability to foresee technological solutions that are possible but do not yet exist, it is almost always more effective and economical to specify the energy or environmental objective rather than a specific means of achieving it. Well designed fiscal incentives provide a clear and consistent signal to the market to make continuing progress toward energy goals. In certain cases they can even correct limitations of the marketplace.

Performance can be measured in different ways: e.g., energy use, petroleum consumption, or GHG emissions.[1] The choice could be important because some fuels that would help reduce petroleum dependence (e.g., coal-derived gasoline) would increase full fuel cycle GHG emissions unless by-product carbon dioxide (CO_2) were captured and sequestered. Choosing GHG emissions as a metric however, would substantially benefit energy security since the preponderance of measures for reducing GHG emissions from transportation vehicles will also reduce oil dependence.

Chapter 7- Good morning, Chairman Baucus, Ranking Member Grassley, and Members of the Committee. My name is John M. Urbanchuk. I am a Director at LECG LLC, a global expert services consulting firm, where I specialize in agriculture and the economics of alternative fuels with particular emphasis on biofuels. I am pleased to be here this morning to discuss the role of the tax code in energy policy and whether technology-neutral energy incentives, including in the areas of fuels, vehicles, electricity, and efficiency, should be developed.

Chapter 8- The following list of authoritative resources is designed to assist in responding to a broad range of constituent questions and concerns about renewable energy and energy efficiency tax incentives. Links are provided for the following: the full text of public laws establishing and extending federal renewable energy and energy efficiency incentives; federal, state, and local incentives resources; incentive resources grouped by technology type (solar, wind, geothermal, and biomass); CRS reports on this topic; and federal grants information resources. The last section of this chapter includes tables displaying popular incentives, the corresponding *U.S. Code* citations, and current expiration dates of those incentives.

This list reflects information that is currently available.

Chapter 9- Historically, U.S. federal energy tax policy promoted the supply of oil and gas. However, the 1970s witnessed (1) a significant cutback in the oil and gas industry's tax preferences, (2) the imposition of new excise taxes on oil, and (3) the introduction of numerous tax preferences for energy conservation, the development of alternative fuels, and the commercialization of the technologies for producing these fuels (renewables such as solar, wind, and biomass, and nonconventional fossil fuels such as shale oil and coalbed methane). The Reagan Administration, using a free-market approach, advocated repeal of the windfall profit tax on oil and the repeal or phase-out of most energy tax preferences — for oil and gas, as well as alternative fuels. Due to the combined effects of the Economic Recovery Tax Act and the energy tax subsidies that had not been repealed, which together created negative effective tax rates in some cases, the actual energy tax policy differed from the stated policy. The George H. W. Bush and Bill Clinton years witnessed a return to a much more activist energy tax policy, with an emphasis on energy conservation and alternative fuels. While the original aim was to reduce demand for imported oil, energy tax policy was also increasingly viewed as a tool for achieving environmental and fiscal objectives. The Clinton Administration's energy tax policy emphasized the environmental benefits of reducing greenhouse gases and global climate change, but it will also be remembered for its failed proposal to enact a broadly based energy tax on Btus (British thermal units) and its 1993 across-the-board increase in motor fuels taxes of 4.3¢/gallon.

In: The Role of Taxes in Energy Production and Conservation ISBN: 978-1-61728-886-9
Editor: Bernadette M. Horton

Chapter 1

ENERGY TAX POLICY: ISSUES IN THE 111TH CONGRESS*

Donald J. Marples and Molly F. Sherlock

SUMMARY

Energy tax policy involves the use of one of the government's main fiscal instruments, taxes (both as an incentive and as a disincentive) to alter the allocation or configuration of energy resources and their use. In theory, energy taxes and subsidies, like tax policy instruments in general, are intended either to correct a problem or distortion in the energy markets or to achieve some economic (efficiency, equity, or even macroeconomic) objective. In practice, however, energy tax policy in the United States is made in a political setting, being determined by the views and interests of the key players in this setting: politicians, special interest groups, bureaucrats, academic scholars, and fiscal dictates. As a result, enacted tax policy embodies compromises between economic and political goals, which could either mitigate or compound existing distortions.

The economic rationale for government intervention in energy markets is commonly based on the government's perceived ability to correct for market failures. Market failures, such as externalities, principal-agent problems, and informational asymmetries, result in an economically inefficient allocation of resources—in which society does not maximize well-being. To correct for these market failures governments can utilize several policy options, including taxes and regulation, in an effort to achieve policy goals.

Current energy policy reflects efforts to achieve both current and past policy objectives. Recent legislative efforts have primarily focused on renewable energy production and conservation to address environmental concerns. In contrast, past efforts attempted to reduce reliance on foreign energy sources through increased domestic production of fossil fuels. Recently enacted legislation focusing on encouraging renewable energy production and conservation reduces reliance on imported, foreign oil, while also addressing environmental

---

* This is an edited, reformatted and augmented edition of a United States Congressional Research Service publication, Report R40999, dated March 8, 2010.

concerns by reducing the use of fossil fuels. Favorable tax preferences given to domestic fossil fuel energy sources also promote domestic energy production, reducing the demand for imported oil.

In the 111th Congress energy tax legislation continues to be focused on increasing incentives for renewable energy production and conservation. The American Recovery and Reinvestment Act of 2009 (ARRA; P.L. 111-5) expanded and extended a number of tax incentives designed to promote renewable energy, conservation, and alternative technology vehicles. In addition, the President's 2010 and 2011 Budget Proposals contain a number of provisions that would, if enacted, continue to move energy policy toward promoting alternative energy sources by eliminating a number of the tax subsidies currently available to the oil and gas industry while also imposing additional taxes that would be borne—at least partially—by these industries. The President's 2011 Budget Proposal continued this policy direction through the proposed removal of a number of subsidies currently available to the coal industry.

INTRODUCTION

Energy tax policy involves the use of one of the government's main fiscal instruments, taxes (both as an incentive and as a disincentive) to alter the allocation or configuration of energy resources and their use. In theory, energy taxes and subsidies, like tax policy instruments in general, are intended either to correct a problem or distortion in the energy markets or to achieve some economic (efficiency, equity, or even macroeconomic) objective. In practice, however, energy tax policy in the United States is made in a political setting, being determined by the views and interests of the key players in this setting: politicians, special interest groups, bureaucrats, academic scholars, and fiscal dictates. As a result, enacted tax policy embodies compromises between economic and political goals, which could either mitigate or compound existing distortions.

The idea of applying tax policy instruments to energy markets is not new, but until the 1970s, energy tax policy had been little used, except to promote domestic fossil fuel production. Recurrent energy-related problems since the 1970s—oil embargoes, oil price and supply shocks, wide petroleum price variations and price spikes, large geographical price disparities, tight energy supplies, and rising oil import dependence, as well as increased concern for the environment— have caused policymakers to look toward energy taxes and subsidies with greater frequency. The direction of U.S. energy tax policy has changed several times since the 197 0s, but appears currently focused on promoting energy efficiency and renewable energy production while continuing to promote U.S. energy security.

This chapter begins with an overview of the economic rationale for intervention in energy markets. Energy markets are uniquely prone to a number of market failures. The nature of these market failures is reviewed, followed by an analysis of interventions with the potential to effectively address these market failures. After presenting the potential energy-related market failures and policy interventions, the current status of U.S. energy tax policy is examined.

U.S energy tax policy as it presently stands aims to address concerns regarding the environment as well as those surrounding national security. Incentives promoting renewable energy production, energy conservation, and alternative technology vehicles address both environmental and national security concerns. Tax incentives for the domestic production of fossil fuels also promote energy security by reducing the nation's reliance on foreign, imported energy sources.

The primary vehicle for energy tax legislation in the 111th Congress to date has been the American Recovery and Reinvestment Act of 2009 (ARRA; P.L. 111-5). This legislation contained a number of provisions that expanded and extended incentives for renewable energy, energy conservation, and alternative technology vehicles. Additionally, the President's Fiscal Year 2010 and 2011 budget proposals contain provisions that would scale back existing incentives for energy production from fossil fuels, and if adopted would decrease the cost of alternative energy relative to energy produced from conventional fossil fuel (coal, oil, and gas) sources. Finally, the American Clean Energy and Security Act of 2009 (H.R. 2454), and several other bills, would control greenhouse gas (GHG) emissions through a cap-and-trade system.

This chapter concludes with a review of major energy tax legislation enacted in the 110th Congress. Legislation enacted in the 110th Congress shifted the focus of energy tax policy, emphasizing alternative and renewable fuels as well as conservation while removing subsidies for oil and gas industry. The Appendix of this chapter provides a brief summary of energy tax policies enacted in the 108th and 109th Congresses.

ECONOMIC RATIONALE FOR INTERVENTION IN ENERGY MARKETS

The primary goal of taxes in the U.S. economy is to raise revenues. There are times, however, when tax policy can be used to achieve other goals. These include the use of tax policy as an economic stimulus or the use of tax policy to achieve social objectives. Tax policy can also be used to correct for market failures, which without intervention result in market inefficiencies. There are a number of market failures surrounding the production and consumption of energy. Tax policy, as it relates to energy, can be used to address these market failures.

Rationale for Intervention in Energy Markets

There are a variety of circumstances in which government intervention in energy markets may improve market outcomes. Generally, government intervention has the potential to improve market outcomes when there are likely to be market failures. Externalities represent one of the most important market failures in energy's production and consumption. Market failures in energy markets also arise from principal-agent problems and information failures. Concerns regarding national security are used to rationalize intervention in energy markets as well.

Externalities

An externality is a spillover from an economic transaction to a third party, one not directly involved in the transaction itself. Externalities are often present in energy markets as both the production and consumption of energy often involve external costs (or benefits) not taken into account by those involved in the energy-related transaction. Instead, these externalities are imposed on an unaffiliated third party. In the presence of externalities, the market outcome will likely lead to an economically inefficient level of production or consumption.

When externalities are present, markets fail to establish energy prices equal to the social marginal cost of supply. The result is a system where cost and/or price signals are inaccurate, such that the socially optimal level of output, or allocative efficiency, is not achieved. Economic theory suggests that a tax be imposed on activities associated with external costs, while activities associated with external benefits be subsidized—in order to equate the social and private marginal costs. These taxes and/or subsidies will result in a more efficient allocation of resources.

Many energy production and consumption activities result in negative externalities, perhaps the most recognized being environmental damage. Air pollution results from mining activities as well as from the transportation, refining, and industrial and consumer use of oil, gas, and coal. Industrial activity can also produce effluents that contaminate water supplies as well as result in other damages to the land. In addition to causing environmental damage, the production and consumption of energy can also lead to lung damage and a variety of other health problems. The use of fossil fuels, both in the production of energy (i.e., coal-fired power plants) and at the consumer level (i.e., using gasoline to power automobiles), and the associated greenhouse gas emissions have contributed to global climate change.[1]

There may also be market failures associated with external benefits stemming from the process of learning-by-doing. Learning-by-doing refers to the tendency for production costs to decline with experience. As firms become more experienced in the manufacturing and use of energy-efficient technologies their knowledge may spill over to other firms without compensation. In energy markets, early adopters of energy-efficient technologies and practices may not be fully compensated for the value of the knowledge they generate.[2]

Principal-Agent and Informational Inefficiencies

Market failures in energy use may also arise due to the principal-agent problem.[3] Generally, the principal-agent problem exists when one party, the agent, undertakes activities on the behalf of another party, the principal. When the incentives of the agent differ from those of the principal, the agent's activities are not undertaken in a way that is consistent with the principal's best interest. The result is an inefficient outcome. In energy markets, the principal-agent problem commonly arises when one party is responsible for making equipment purchasing choices while another party is responsible for paying the energy costs, which are related to the efficiency level of the purchased equipment.

For residential rental properties, the incentives for the landlords and tenants surrounding the adoption of energy-savings practices are often not aligned. Landlords will under-invest in energy- saving technologies for rental housing when the benefits from such investments accrue to tenants (i.e., tenants are responsible for paying their own utilities) and the landlord does not believe the costs of installing energy-saving devices can be recouped via higher rents. Tenants do not have an incentive to invest in energy-savings technologies in rental units when their expected tenure in a specific property is relatively short, and they will not have enough time to reap the full benefits of the energy conserving investments. There is also evidence that when utilities are included in the rent, tenants do not engage in energy conserving behaviors. On the other hand, when tenants pay utilities on their own, energy-saving practices are more frequently adopted.[4] The implication is that inefficient energy use by tenants in apartments where utilities are included as part of the rent would offset energy-saving investments made by landlords; consequently, landlords under-invest in energy efficiency. In general, the under-investment in energy conservation measures in rental housing provides economic rationale for intervention.

In another example, the incentives of homebuilders and homebuyers may not be aligned. Consequently, the principal-agent problem may result in an inefficient utilization of energy-efficient products in newly constructed homes. Homebuilders may have an incentive to install relatively low efficiency products to keep the cost of construction down if they do not believe that the cost of installing energy-efficient products will be recovered upon sale of the property. The value of installing energy-efficient devices may not be recoverable if builders are not able to effectively communicate the value of energy-efficient devices once installed. Further, since homebuilders are not able to observe the energy use level of prospective buyers they may not be able to choose the products that best match the use patterns of the ultimate energy consumer. The result may be less energy efficiency in new homes.

There are also informational problems that may lead to underinvestment in energy-efficient technologies. For example, homeowners may not know the precise payback or rate of return of a specific energy-efficient device. This may explain the so-called "energy paradox"—the empirical observation that consumers require an abnormally high rate of return to undertake energy- efficiency investments.[5]

National Security

Preserving national security is another often cited rationale for intervention in energy markets. Presently, much of the petroleum consumed in the United States is derived from foreign sources. There are potentially a number of external costs associated with petroleum importation, especially when imported from unstable countries and regions. First, a high level of reliance on imported oil may contribute to a weakened system of national defense or contribute to military vulnerability in the event of an oil embargo or other supply disruption. Second, there are costs to allocating more resources to national defense than necessary when relying on high levels of imported oil. Specifically, there is an opportunity cost associated with resources allocated to national defense, as such resources are not available for other domestic policy initiatives and programs. To the extent that petroleum importers fail to take these external costs into account, there is market failure.

In addition, the economic well-being and economic security of the nation depends on having stable energy sources. There are economic costs associated with unstable energy supplies. Specifically, increasing unemployment and inflation may follow oil price spikes.[6]

Potential Interventions in Energy Markets

When there are negative externalities associated with an activity, correcting the economic distortion with a tax, if done correctly, can improve economic efficiency.[7] Conversely, when there are positive externalities associated with an activity, a subsidy can improve economic efficiency. The tax (subsidy) should be set equal to the monetary value of the damages (benefits) to third parties imposed by the activity.[8] The tax serves to increase the price of the activity, and reduce the equilibrium quantity of the activity, while a subsidy reduces the price, increasing the equilibrium quantity of the activity.

The production and consumption of fossil fuel energy can have negative externalities via detrimental environmental impacts. While multiple policy options to address this externality exist, economists tend to favor an emissions tax to address this externality because of such a tax's efficiency advantage.[9] Current political momentum, however, is behind a cap-and-trade policy, as is discussed below.

An alternative approach to reducing the use of fossil fuels has been to subsidize energy production from alternative energy sources. There are concerns, however, that using subsidies to stimulate demand for clean alternative fuels, as opposed to fossil fuels, may not be economically efficient. First, subsidies reduce the average cost of energy, and as the average cost of energy falls, the quantity of energy demanded increases, countering energy conservation initiatives.[10] Second, while the subsidy is intended to enhance economic efficiency, subsidies may be inefficient to the extent they are funded using distortionary taxes.[11] Hence, the more efficient alternative may be to place a tax on the undesirable activity.

The tax code subsidizes research and development (R&D) because of the associated positive externalities generated. There is the potential for learning-by-doing from early adopters of energy-efficient technologies, indicating that there may be positive external effects associated with these activities. For this reason, subsidies given to early adopters may enhance economic efficiency. Further, positive externalities are associated with R&D activities that lead directly to technological innovations.[12] In addition to budgeted spending on R&D, the tax code provides incentives for firms to engage in energy R&D (for example, the energy research credit (IRC §41)).[13]

When principal-agent problems lead to a market failure, economically efficient corrective measures would be those that increase the equilibrium quantity of the underprovided good (energy-efficient technologies in the present context). Currently, the definition of a taxpayer's gross income excludes any subsidy provided by a public utility to a consumer for the purchase or installation of energy-saving devices (see IRC § 136). This exclusion subsidizes energy-efficient devices. This exclusion does not specifically target the inefficiency in rental housing created by the principal-agent problem, since the exclusion applies to both owner- and non-owner-occupied property. Nonetheless, the exclusion may serve to ameliorate the market failure in rental property.

There are also various options for market intervention to address the informational problem associated with energy consumption and energy-efficient technologies. One option would be an information-based solution, such as energy-efficiency labeling and education and awareness campaigns. Alternatively, a tax-incentive-based approach—such as a credit or deduction for the purchase of energy-efficient devices—could be used to address the market inefficiency. Given that this market failure is an informational problem, it might be more efficient to pursue information-based solutions (such as energy-efficiency labeling like the U.S. Environmental Protection Agency's Energy Star program).

Table 1. Energy Tax Expenditures
billions of dollars

Tax Expenditure	Description	Cost 2008-2012	Expiration Date	I.R.C. Section
Fossil Fuels				
Expensing of percentage over cost depletion	Firms that extract oil or gas are permitted to deduct 15% of sales (up to 25% for marginal wells depending on oil prices) to recover their capital investment in a mineral reserve.	$7.9	none	611, 612, 613, 613A, 291
Expensing of exploration and development costs	Firms engaged in the exploration and development of oil, gas, or geothermal properties have the option of expensing (deducting in the year paid or incurred) rather than capitalizing (i.e., recovering such costs through depletion or depreciation) certain intangible drilling and development costs (IDCs).	$7.5	none	263(c), 291, 616-617, 57(a)(2), 59(e), 1254
Amortization of G&G expenditures associated with oil and gas exploration	Under the Modified Accelerated Cost Recovery System (MACRS), the cost of selected types of geological and geophysical property is depreciated over 2 years for independent producers.	$1.3	none	167(h)
Credit for enhanced oil recovery costs	A 15% income tax credit for the costs of recovering domestic oil by qualified "enhanced-oil-recovery" (EOR) methods. Other costs associated with tertiary injectants are also deductible.	$0.2	none	43, 193
Coal Production Credits	Production credit of $6.20-per-ton production credit for refined coal used to produce steam, or $2 per barrel-of-oil equivalent credit for refined coal used as a steel industry fuel, or a $1.625 per-ton production credit (all adjusted for inflation from 1992) for coal reserves owned by an Indian tribe.	$0.2	12/31/2009 (refined coal) 12/31/2012 (Indian coal)	45
Credits for investing in clean coal facilities	Tax credit of 20% of investment for integrated gasification combined cycle (IGCC) systems; 15% for other advanced coal technologies	$0.8	none (credit allocation limit)	48A
Amortization of air and pollution control facilities	Allows the pre-1976 5-year amortization period for investments in pollution control equipment for coal-fired electric generation plants available to those plants placed in service on or after January 1, 1976. The 5-year amortization	$0.6	none	169

Table 1. (Continued)

Tax Expenditure	Description	Cost 2008-2012	Expiration Date	I.R.C. Section
	incentive for pre-1976 plants applies only to pollution control equipment with a useful life of 15 years or less. In that case 100% of the cost can be amortized over five years. If the property or equipment has a useful life greater than 15 years, then the proportion of the costs that can be amortized over five years is less than 100%.			
	Renewable Energy Production			
Credits for elect-ricity production from renewable resources	Tax credit of 2.1¢/kWh for electricity produced from wind, closed-loop bio-mass, geothermal energy, and solar sources. Tax credit of 1.1¢/kWh for elec-tricity produced from open-loop biomass, small irrigation, landfill gas, trash combus-tion, qualified hydropower, marine and hydrokinetic sources. The tax credit is given for 10 years after the date the facility is placed in service.	$7.1	Property must be placed in ser-vice by 12/31/13 (12/31/12 for wind)	45
Energy credit	Tax credit equal to 10% of investment in energy production using geothermal, microturbine, ground water for heating or cooling, or combined heat and power methods. The tax credit is equal to 30% of investment in energy production using solar electric, solar hot water, fuel cell or small wind methods.	(i)	none (geother-mal, solar elect-ric, solar hot water and small wind electric) 12/31/2016 (else)	48
Residential energy-efficient property credit	Tax credit for 30% of the cost of the purchase of solar electric property, solar water heating property, geothermal heat pump property, or small wind energy property. Fuel cell power plants receive 30% credit, limited to $500 for each 0.5 kilowatt of capacity.	$0.1	12/31/2016	25D
Five-year cost recovery of certain energy property	Accelerated depreciation allowances are provided under the modified accelerated cost recovery system (MARCs) for investments in certain energy property. Specifically, certain solar, wind, fuel cell, microturbine and biomass property has a five year recovery period. Cellulosic biofuel plant property is allowed an additional first-year depreciation deduction equal to 50% of the property's adjusted basis.	$0.5	Cellulosic bio-fuel property must be placed in service by 12/31/2012	168
Credits for holders of clean renewable energy bonds	Provides a tax credit for the holder of the bond against its income tax. Clean Renewable Energy Bonds ("CREBs") are subject to a volume cap of $1.2 billion with a credit rate set to allow the bond to be issued at par and without interest.	$0.5	volume limited	54, 54C

Table 1. (Continued)

Tax Expenditure	Description	Cost 2008-2012	Expiration Date	I.R.C. Section
	New Clean Renewable Energy Bonds ("New CREBs") are subject to a volume cap of $2.4 billion with a credit rate set at 70% of what would permit the bind to be issued at par and without interest.			
Credit for alcohol fuels	Income or excise tax credit. Ethanol tax credit generally 45¢ per gallon (extra 10¢ for small producers; alcohol tax credit generally 60¢ per gallon; $1 per gallon for biodiesel, agribiodiesel, and renewable diesel (extra 10¢ for small producers of agribiodiesel) alternative fuels generally 50¢ per gallon; cellulosic biofuels generally $1.01 per gallon.	$0.3 (ii)	12/31/09 for biodiesel, agribiodiesel, renewable biodiesel, black liquor, and alternative fuels; 12/31/10 for alcohol fuels; 12/31/12 for cellulosic biofuel	40, 40A, 6426, 6427(e)
Energy Conservation				
Credit for nonbusiness energy property	Tax credit for 30% of the amount paid for qualified energy-efficiency improvements and expenditures for residential energy property including qualifying improvements to the building's envelope, the HVAC system, furnaces, or boilers. Credit limited to $1,500 for 2009 and 2010 combined.	$2.1	12/31/2010	25C
Deduction for expenditures on energy-efficient commercial property	Tax deduction for the cost of building envelope components, heating cooling systems, and lighting. The deduction is limited to $1.80 per square foot.	$0.2	12/31/2019	179D
Exclusion of energy conservation subsidies provided by public utilities	Subsidies are not taxable as income.	$0.1	none	136
Energy-efficient home credit	Contractor may claim $1,000 credit for building homes 30% more efficient than the standard; $2,000 credit for homes 50% more efficient than the standard	$0.1	12/31/2009	45L

Table 1. (Continued)

Tax Expenditure	Description	Cost 2008-2012	Expiration Date	I.R.C. Section
Credit for producing energy-efficient appliances	Tax credit based on energy efficiency, energy savings. Maximum credit is $75 for dishwashers, $200 for refrigerators, and $250 for clothes washers.	$0.3	12/31/2010 (earlier expir-ation for lower efficiency standards)	45M
Qualified energy conservation bonds	Federal government has authorized the issue of $3.2 billion in Qualified Energy Conservation ("QECBs"). QECBs provide a tax credit worth 70% of the tax credit bond rate stipulated by the Secretary of the Treasury. QEC bonds issued by state and local governments must fund an energy-savings project, such as the green renovation of a public building, R&D in alternative fuels, public transportation projects or public education campaigns.	$0.1	volume limited	54D
Alternative Technology Vehicles				
Hybrid vehicles	The first 60,000 hybrid cars or light trucks sold per manufacturer are eligible for a credit of $400 to $2,400 (depending on fuel economy). An additional credit of $250 to $1,000 is available depending on a vehicles expected lifetime fuel savings. Heavy vehicles (those exceeding 8,500 pounds) qualify for up to $30,000 in credits and is not subject to a volume cap.	$0.9	12/31/09 (12/31/10 for vehicles weig-hing more than 8,500 pounds)	30B
Other alternative fuel vehicles	Fuel cell vehicles receive a base credit of $8,000 (reduced to $4,000 after 2009) for vehicles weighing less than 8,500 pounds. Heavier vehicles qualify for up to a $40,000 credit. An additional credit of up to $4,000 is available for cars and light trucks that exceed the 2002 base fuel economy. Lean burn vehicles are eligible for the same credit as hybrid vehicles. Alternative fuel vehicles can qualify for a credit of up to $5,000 for cars and light trucks and $40,000 for heavy vehicles. Credit amount varies according to the vehicle's incremental cost and ratio of alternative fuel use. Credits available for plug-in electric vehicles are available up to $7,500 (cars and light trucks) depending on kilowatt hour capacity of vehicle and $15,000 for heavy vehicles. A 10% credit, up to $2,500, is available for the cost of electric-drive low-speed neighborhood vehicle, motorcycle and three-wheeled vehicles.	$0.1	12/31/2010 for advanced lean burn vehicles, and alternative fuel vehicles; 12/31/11 for ele-ctric drive low speed vehicles and conv-ersion to plugin vehicle 12/31/2014 for fuel cell vehicles.	30, 30B, 30D

Table 1. (Continued)

Tax Expenditure	Description	Cost 2008-2012	Expiration Date	I.R.C. Section
	A 10% credit, up to $4000, is available for conversion to a plug-in electric drive vehicle.		12/31/2009 for plug-in low sp-eed vehicles and heavy vehicles. Credit for plug-in vehicle volume capped.	
Other / Miscellaneous				
Election to expense 50% of qualified property used to refine liquid fuels	A taxpayer may elect to expense 50% of the cost of any qualified property used for processing liquid fuel from crude oil or qualified fuels. The remainder is recovered under the otherwise applicable rules.	$3.8	12/31/2013	179(c)
Exceptions for ene-rgy related Publicly Traded Partnerships	Publicly Traded Partnerships are generally treated as corporations. The exception from this rule occurs if at least 90 percent of its gross income is derived from interest, dividends, real property rents, or certain other types of qualifying income. Qualifying income includes income derived from certain energy-related activities.	$2.6	none	7704, 851
Exclusion of inter-est on State and local government private activity bonds for energy produc-tion facilities	Exclusion of interest from private activity bonds used to finance privately ow-ned and/or operated sewage, water, solid waste disposal, and heating and coo-ling facilities, certain private electric and gas facilities, hydroelectric dam en-hancements, qualified green building and sustainable design projects from tax.	$0.6	none	141, 142
Credit for producing fuels from a non-conventional source	A production tax credit of $3 per barrel of oil-equivalent (in 1979 dollars) for certain types of liquid, gaseous, and solid fuels produced from selected types of alternative energy sources (termed "non-conventional fuels"), and sold to unrelated parties. The full credit is available if oil prices fall below $23.50 per barrel (in 1979 dollars); the credit is phased out as oil prices rise above $23.50 (in 1979 dollars) over a $6 range (i.e., the inflation-adjusted $23.50 plus $6). The phase out limit does not apply to coke or coke gas.	$0.6	placed in service by 12/31/2009 (coke credit)	45K, 40A, 6426

Table 1. (Continued)

Tax Expenditure	Description	Cost 2008-2012	Expiration Date	I.R.C. Section
	In addition, a $1.00 per gallon tax credit is available for both biodiesel and renewable biodiesel production, while a 10¢ per gallon small producer credit is only available on the former. Further, a 50¢ per gallon credit is available on the production of alternative fuels.			
Depreciation recovery periods for energy specific items	Smart electric distribution property is allowed 10-year depreciation under the modified accelerated cost recovery system (MACRs). Certain electric transmission property is allowed a 15-year depreciation. Natural gas distribution lines are also allowed a 15-year depreciation.	$1.3	various	168(e)
Deferral of gains from the sale of electric trans-mission property	A taxpayer may elect to recognize the gain from the sale of certain electric transmission property over an eight year period.	$0.3	12/31/2009	451
Credits for clean fuel vehicle refu-eling property	A 50% credit (capped at $50,000 for business property and $2,000 for nonbusiness property) for 2009 and 2010) and a 30% credit for hydrogen property (capped at $200,000).	$0.1	12/31/10 (12/31/14 for hydro-gen refueling property)	30C

Source: CRS compilation based on Joint Committee on Taxation, *Tax Expenditures for Energy Production and Conservation*, JCX-25-09R, April 21, 2009.

Notes: i- less than $50 million per year; ii- the credit from excise tax for alcohol fuels results in a reduction of excise tax receipts of $13.6 billion from 2008 – 2012. These estimates are based on the January 2008 Congressional Budget Office baseline for tax expenditures under present law as of October 31, 2008. *- indicates that the provision was modified under ARRA. Table 2 provides the additional revenue loss associated with the ARRA modification.

Finally, there are questions regarding the most efficient and effective mode of intervention to address the negative external costs, specifically national and economic security concerns, associated with the consumption of imported oil. One option would be to impose a tax to correct the distortion. There are two problems with imposing such a tax. First, a tax on imported oil is likely to violate trade agreements. This has led policymakers to pursue policies that subsidize domestic petroleum production.[14] The second problem is that oil is a commodity priced on world markets. The United States producing oil for its own use does not necessarily insulate consumers from global fluctuations in oil prices. Additionally, to the extent that oil price fluctuations impact export prices in other parts of the world, such as Europe and China, the United States is still likely to experience economic impacts from oil price fluctuations.[15]

Taxes as a User Charge

Energy taxes may be employed as user charges for a public good or a quasi-public good.[16] In the United States, non-toll highways and highway infrastructure have the public good property of non-excludability. Highways are not likely to be provided by the market because public goods and quasi-public goods are susceptible to the free-rider problem.[17] If the private market fails to provide a public good, like highways, then government intervention via provision of highways can enhance economic efficiency. The federal excise tax on gasoline is often rationalized as a user fee for the federal highway system.[18] For the tax to be efficient and equitable, it would charge individuals in proportion to their benefit from the public good (the highway system). In practice, gas taxes do not reflect the cost to the user but instead depend on the fuel efficiency of a specific vehicle.[19] Furthermore, some of the revenues collected from the federal gas tax serve to subsidize public transportation, undermining the view of the federal gas tax as a highway user fee.

Current Status of U.S. Energy Tax Policy

Current U.S. energy tax policy appears to be aimed at stemming growth in U.S. dependence on imported oil, especially from volatile regions of the world. This reflects the belief that national security is linked to energy security. Many of the specific policies currently in place are the result of past legislative action and initiatives. The short-run policies aim to increase the domestic energy production of fossil fuels, while the long-run policies appear aimed at promoting energy conservation and the use of renewable energy sources. Recently, attention has focused on the transportation sector, in the nature of both technology and alternative fuels. Table 1 contains a current list of energy-related tax expenditures.[20]

Energy conservation measures and domestic production of energy from renewable resources help reduce demand for fossil fuels by diversifying the sources from which energy can be derived to meet U.S. demand. Unlike domestic fossil fuel subsidies, these policies are long-term because they require a commitment in the face of volatility in fossil fuel prices and technologies that may not have reached the stage where they are competitive in the market in the absence of subsidization.

Current energy tax policy is the result of prior policy action undertaken in an effort to achieve the nation's long-standing goal of enhancing U.S. energy security. For example, the promotion of domestic fossil fuel production, the current principle short-run strategy, was a central tenet of energy tax policy from 1918 through the late 1 960s. Further, the current long-run policies of conservation and alternative fuel sources have origins in tax policies from the 1970s. The perceived means of achieving U.S. energy security have shifted over the years—and these shifts are clearly reflected in the current status of U.S. energy tax policy.

Energy tax policy proposed in the Obama Administration's Fiscal Year 2011 Budget Proposal differs substantially from current U.S. energy tax policy. If fully enacted, the FY20 11 Budget Proposal would reduce or eliminate several energy tax policies that encourage energy production from the coal, oil, and gas industries, while expanding incentives for conservation and renewable energy production. This reflects the shift in energy tax policy from one primarily focused on enhancing U.S. energy security through diversification of energy resources towards a tax policy that more readily incorporates environmental concerns (tax policy that discourages the use of fossil fuels, regardless of their nation of origin).

Energy tax policy—like all tax policy—can lead to unintended consequences. Notably, this issue has arisen in the 111th Congress in its deliberations concerning "black liquor." In the context of taxes, the term "black liquor" currently refers to a process in which pulp mills use a mixture of conventional fuel and a byproduct of the pulping process as an energy source for the mill. According to changes enacted in The Safe, Accountable, Flexible, Efficient Transportation Equity Act: A Legacy for Users (P.L. 109-59; SAFETEA-LU), "black liquor" is eligible for the alternative fuels tax credit, which was not the congressional intent of the provision.[21] Recognizing the unintended consequence, Senate Finance Committee Chairman Max Baucus[22] stated in response to draft legislation, "Our measure ensures this tax credit is used consistently as the law intended, not through an unintended loophole," and Senate Finance Committee Ranking Member Charles Grassley[23] stated, "The paper industry was not intended to receive the alternative fuels tax credit when the credit was enacted.... This staff draft proposal is meant to clarify that legislative intent."

A further unintended consequence of "black liquor" and selected other energy tax credits—such as those for ethanol—is that by reducing the total cost of blended fuels, they may actually increase the consumption of fossil fuels that the credits were designed to reduce.[24]

Fossil Fuel Production

There are a number of tax incentives currently available for energy production using fossil fuels. They can be broadly categorized as either enhancing capital cost recovery or subsidizing extraction of high-cost fossil fuels. Between 2008 and 2012, the total cost of tax expenditures related to fossil fuels is estimated to be $18.5 billion.

Among the capital cost subsidies, the allowance of the percentage depletion method is estimated to cost the most in foregone revenue, $7.9 billion between 2008 and 2012.[25] Under percentage depletion, a deduction equal to a fixed percentage of the revenue from the sale of a mineral is allowed. Total lifetime deductions, using this method, typically exceed the capital invested in the project. To the extent that percentage depletion deductions exceed project investment, percentage depletion becomes a production subsidy, instead of an investment subsidy. Other capital cost recovery provisions include expensing of intangible drilling costs related to exploration and development and a decrease in the amortization period for certain geological and geophysical property.[26] The expensing of exploration and development costs is also a relatively large tax expenditure, estimated to cost the federal government $7.5 billion in revenue losses over the 2008 through 2012 budget window.

Compared to the capital cost recovery provisions, tax expenditures intended to offset high extraction or refining costs are small. Credits for refined coal production, enhanced oil recovery costs and investing in clean coal facilities collectively are estimated to result in approximately $1.2 billion between 2008 and 2012.

Renewable Energy Production

Several tax incentives subsidize the production of energy from renewable sources. While the specific incentives differ in design, they all work to increase the after-tax return on an investment in renewable energy production by providing tax incentives on the condition of eligible investment or production. Between 2008 and 2012, the total cost of tax expenditures related to the production of renewable energy is estimated to be $8.5 billion.[27]

Among the renewable energy tax expenditures, the credit for electricity production from renewable sources is the most expensive, with an estimated cost of $7.1 billion over the 2008 through 2012 budget window.[28] This production tax credit provides a varying incentive for electricity generation from selected renewable energy sources.[29] Other renewable energy production tax incentives include credits for the production or blending of alcohol fuels and renewable diesel.

In terms of budgetary impact, there are a number of smaller tax expenditures related to renewable energy. First, there is the energy credit, which provides a credit equal to either 10% or 30% of eligible investment in renewable energy production.[30] Second, there is the residential energy- efficient property credit, which provides a tax credit for the installation of renewable electricity generating property for a residential dwelling. Third, the reduced depreciable life for renewable energy investments provides an additional subsidy for businesses. Finally, clean renewable energy bonds (CREBs) and new clean renewable energy bonds (New CREBs) are subsidized in that issuers are not required to pay interest to investors, as investors receive a federal tax credit in lieu of interest payments.[31]

Energy Conservation

Incentives for energy conservation primarily operate by providing incentives to owners of residential and commercial property that undertake energy-efficient upgrades. There are also incentives to manufacturers of energy-efficient appliances and for the issuance of qualified energy conservation bonds. Between 2008 and 2012, the total cost of tax expenditures related to energy conservation is estimated to be $2.9 billion.[32]

The bulk of spending on energy conservation encourages property owners to undertake energy- efficiency improvements on existing buildings. The credit for nonbusiness energy property, which provides homeowners up to a $1,500 tax credit for expenditures on energy-efficient property such as windows, doors, furnaces, and boilers, is the largest energy conservation tax expenditure, with an estimated cost of $2.1 billion during the 2008 through 2012 budget window. In addition, energy-efficient improvements for commercial property, including upgrades to a building's envelope, heating and cooling, or lighting system are eligible for a tax deduction, limited to $1.80 per square foot. Finally, the exclusion of subsidies provided by utility companies to energy consumers undertaking energy-efficiency upgrades from income increases the value of such subsidies, encouraging individuals to undertake such improvements.

There are two additional tax incentives designed to encourage manufacturers to build energy- efficient products. First, the energy-efficient home credit gives homebuilders up to a $2,000 credit for new energy-efficient homes. This credit expires December 31, 2009. Second, the credit for producing energy-efficient appliances makes manufacturers of energy-efficient dishwashers, refrigerators, and clothes washers eligible for a credit. The cost of these two programs to the federal government is $0.4 billion over the 2008 through 2012 budget window.

Qualified energy conservation bonds (QECBs) also encourage energy conservation, by providing subsidized financing to energy conservation projects. QECBs are tax credit bonds, where the holder of the bond receives a federal tax credit in lieu of interest payments.

Alternative Technology Vehicle Credits

Various credits are currently in place for the purchase alternative technology vehicles.[33] The tax code defines alternative technology vehicles to include qualified fuel cell vehicles, hybrid vehicles, advanced lean burn technology vehicles, and alternative fuel vehicles with credit amounts varying by the specific technology and vehicle type. Between 2008 and 2012, the total cost of tax expenditures related to hybrid and alternative technology vehicles is estimated to be $1.0 billion.[34]

Other Energy Tax Provisions

There are a number of other energy tax provisions that do not fall under the fossil fuel, renewable energy, energy conservation, or alternative vehicle technology categories. The largest of these tax expenditures is the provision allowing taxpayers to expense 50% of the cost of property used to process qualified fuels (with an estimated budget cost of $3.8 billion over the 2008 through 2012 budget window). The remaining provisions include those giving special tax treatment for energy related Publicly Traded Partnerships, accelerated depreciation for various energy specific items, excluding interest from private activity bonds related to energy production, deferral of gains from the sale of electric transmission property and credits for installing clean fuel vehicle refueling property. Between 2008 and 2012, the total cost of tax expenditures related to other energy tax provisions is estimated to be $9.3 billion.[35]

Table 2. Energy Tax Provisions Enacted under American Recovery and Reinvestment Act of 2009 Billions of Dollars

Tax Expenditure	Changes Under ARRA	Cost 2009-2013	I.R.C. Section
	Renewable Energy Production		
Extension of the renewable energy production tax credit	Extended the placed in service dates for wind through December 31, 2012 and December 31, 2013 for other facilities.	2.9	45
Election of investment credit in lieu of the production credit	Businesses that place in service facilities producing electricity from a renewable resource can elect to claim an investment credit in lieu of the production credit.	0.3	48
Modify energy credit	Remove cap for wind systems and remove cutback to credit for subsidized energy financing.	0.2	48
Credit for residential energy-efficient property	Increased the credit rate to 30% and removed credit limits.	0.1	25D
Increase limitation on new clean renewable energy bonds	Increased funds available for the issue of new clean renewable energy bonds.	0.1	54D
	Energy Conservation		
Extend and increase credit for nonbusiness energy property	Removed maximum amounts previously associated with credit and increased the cre-dit rate to 30% up to at $1,500 maximum.	1.9	25C
Increase limitation on qualified energy conservation bonds	Increased funds available for the issue of qualified energy conservation bonds.	0.1	136
	Alternative Technology Vehicles		
Modify alternative motor vehicle credit and credit for plug-in electric vehicles	Modified existing credit and added credit for low-speed two and three wheeled vehicles.	0.1	30D
Credit for plug-in vehicle conversion	Introduced credit for plug-in electric drive conversion kits. Credit is 10% of expenditures limited to $4,000. The credit expires on January 1, 2012.	(i)	30B
	Other		
Changes to credit for refueling and recharging property	Increased credit rate from 30% to 50% and limits from $30,000 to $50,000 for business property and $1,000 to $2,000 for nonbusiness property; added $200,000 limit for hydrogen property.	0.1	30C
Five-year recovery period for depreciation of smart meters	Reduced recovery period for smart meters from 10 years to five years for property placed in service before January 1, 2011.	0.1	168(e)
Research credit for qualified energy research	Added a 20% credit for qualified expenses on energy research during 2009 and 2010.	(i)	41

Source: CRS and Joint Committee on Taxation, JCX-18-09

Notes: The revenue losses are losses associated with the provisions as enacted under ARRA.

Energy Tax Legislation in the 111th Congress

Energy tax legislation in the 111th Congress has continued to focus on increasing incentives for renewable energy production and energy conservation. Further, recent legislation illustrates how the tax system can attempt to encourage specific behavior, while discouraging other behavior, in a quest to achieve a policy goal. The American Recovery and Reinvestment Act of 2009 (ARRA; P.L. 111-5) utilizes incentives to encourage the production of energy from renewable sources and promote energy conservation. In contrast, the President's Fiscal Year 2010 and 2011 Budget Proposals aim to promote renewable energy through the elimination of several incentives for conventional energy. As the 111th Congress progresses, it is expected that the focus of energy tax policy will continue to be on the promotion of renewable energy production and energy conservation.

The American Recovery and Reinvestment Act of 2009 (P.L. 111-5)

From the perspective of energy tax policy, ARRA modified incentives for renewable energy production, energy conservation, alternative technology vehicles, as well as a number of other energy tax incentives.[36] Taken together, these incentives are estimated to reduce federal revenue collections by nearly $6 billion over the five-year 2009 through 2013 budget window and nearly $20 billion over the 2009-2019 budget window. Table 2 categorizes and provides details on the specific provisions contained within ARRA. Collectively, ARRA's energy tax provisions lower the cost of selected renewable energy relative to energy from other sources, such as oil and gas.

Provisions enacted under ARRA extended and expanded a number of incentives to encourage the production of energy using renewable resources. Specifically, the renewable energy production tax credit (PTC) was extended through 2013.[37] The PTC provides a credit per kilowatt-hour of electricity produced by a variety of qualified renewable resources.[38] In addition, provisions in ARRA temporarily allow firms to claim the business energy investment tax credit in lieu of the PTC. Additionally, ARRA allowed projects eligible for a PTC or investment tax credit to substitute a one time grant from the U.S. Treasury. Allowing investors to take a one time grant instead of future tax credits addresses uncertainty renewable energy investors may have regarding future tax positions.

The expansion and extension of the energy credit and the credit for residential energy-efficient property also promote renewable energy production. Under ARRA, caps previously associated with certain energy credits were removed. Businesses can now claim an uncapped 30% tax credit for qualified small wind energy property.[39] In addition, individuals are now allowed to claim an uncapped 30% tax credit for qualified solar heating property, small wind energy property, and geothermal heat pumps.[40] Renewable energy production was also further encouraged by ARRA's provision increasing the funds available for the issue of new clean renewable energy bonds.

ARRA contained two tax provisions specifically encouraging energy conservation. The first provision modified the tax credits for energy-efficient improvements to existing homes by temporarily increasing the credit rate and removing credit caps previously associated with specific types of property. For qualified energy-efficiency improvements, such as the installation of energy-efficient building envelope components, furnaces, or boilers, installed during 2009 and 2010, taxpayers can claim a 30% tax.[41] ARRA also removed property-by-property caps on the tax credit and replaced them with a $1,500 cap for the total amount of the credit claimed.[42] The second energy conservation provision increased funds available for the issue of qualified energy conservation bonds.

To further promote alternative technology vehicles, tax provisions enacted under ARRA modified the credits for alternative fuel vehicles and plug-in electric vehicles. Additionally, a tax credit for plug-in vehicle conversion was introduced.

There were also a number of other energy tax provisions in ARRA. First, the credit rates and limits for refueling and recharging property were increased. Second, the recovery period for depreciating smart meters was temporarily decreased from 10 years to five years. Third, a 20% tax credit was introduced for expenses related to qualified energy research.

The total cost of these energy tax provisions enacted under ARRA is estimated to be $5.9 billion over the 2009 through 2013 budget window. The majority of these revenue losses, $3.6 billion, are for incentives promoting energy production from renewable sources. Incentives to encourage energy conservation had revenue losses of $2 billion over the 2009 through 2013 budget window. Modification to the incentives for alternative technology vehicles are associated with $0.1 billion in revenue losses over the 2009 through 2013 budget window, while other energy tax provisions are associated with revenue losses of $0.2 billion over the same time period.

The President's Fiscal Year 2010 and 2011 Budget Proposals

A reduction in or the repeal of selected tax preferences for the oil and gas industry appears to be the focus of energy tax policy in the President's Fiscal Year 2010 and 2011 Budget Proposals. Specifically, the FY2010 and FY 2011 Budget Proposals contain several provisions directly targeting the oil and gas industries. These provisions include the domestic manufacturing deduction for oil and gas production, the repeal of percentage depletion, and the expensing of intangible drilling costs—three provisions which have been scaled back in recent years—along with the repeal of the deduction for tertiary injectants and credits for enhanced oil recovery projects and production from marginal wells. In addition, the FY20 10 and FY2011 Budget Proposals would levy an excise tax on oil and gas production on the Outer Continental Shelf and increase the amortization period for geological and geophysical costs from five to seven years.

The FY2010 and FY2011 Budget Proposals also contains several additional proposals that would be borne—at least partially—by the oil and gas industry. The Budget Proposals would reinstate the superfund excise taxes and environmental income tax.[43] Notable among the Superfund taxes is the excise tax on domestic crude oil and on imported petroleum products, which is taxed at a rate of 9.7 cents per barrel. Further the FY20 10 and FY20 11 Budget Proposals would repeal the last-in, first-out (LIFO) method of inventory accounting. This method of inventory accounting values the goods sold as the most recent inventory purchase. During a period of rising prices, this method of inventory accounting increases production costs and reduces taxable income and tax liabilities.

In addition to items common to both Budget Proposals, the FY20 11 Budget Proposal also contains several proposals targeted at the coal industry. The elimination of percentage depletion for fossil fuels is estimated to raise nearly $1.1 billion over the 2011-2020 budget window and represents nearly one-half of the projected revenue gain from repeal of the coal industry tax preferences identified in the FY2010 Budget Proposal.

Taken as a whole, these provisions raise the cost of coal, oil, and gas production and can be viewed as reducing the cost of alternative energy production relative to energy production from conventional energy sources. That is, if enacted, the FY 2010 and FY2011 Budget Proposal would decrease the relative cost of alternative energy production by increasing the cost of coal, oil, and gas production.[44]

American Energy Production and Price Reduction Act (H.R. 3505)

This act proposes to increase energy production domestically. Specifically, the act seeks to increase American oil production, encourage the use of nuclear and solar energy production, as well as provide incentives for natural gas and clean coal. Additionally, the act would expand a number of existing credits promoting energy efficiency.

Carbon Tax / Climate Change[45]

As of the date of this chapter, Members in the 111th Congress have introduced seven stand-alone proposals that would control greenhouse gas (GHG) emissions.[46] The proposals offered to date would employ market-based approaches—either a cap-and-trade or carbon tax system, or some combination thereof—to reduce GHG emissions. The legislative proposals are varied in their overall approaches in controlling GHG emissions. Some control emissions by setting a quantity (or cap); others control emissions by setting a price (or tax/fee). In addition, the proposals differ in their inclusion of particular design elements, such as whether or not to allow offsets (emission reduction opportunities from economic sectors not directly addressed by the primary approach).

Three of the proposals—H.R. 594 (Stark), H.R. 1337 (Larson), and H.R. 2380 (Inglis)—would use a carbon tax approach to address carbon dioxide (CO2) emissions from fossil fuel combustion. H.R. 1683 (McDermott) would establish a program that may be described as a dynamic carbon tax: the tax rate would be linked with annual emission allocations (or caps).

These carbon tax proposals, however, are not widely seen as the primary energy and climate change legislative proposal in the 111th Congress. Instead, cap-and-trade programs have constituted the primary energy and climate change legislative proposals in the 111th Congress. Current cap-and-trade legislative vehicles include H.R. 2454 and S. 1733. The American Clean Energy and Security Act of 2009 (H.R. 2454) (Waxman/Markey) which has passed the House would establish cap-and-trade programs.[47] The Clean Energy Jobs and American Power Act (S. 1733), was considered and ordered to be reported by the Committee on the Environment and Public Works, would also rely on a cap-and-trade program to reduce greenhouse gas emissions.[48] H.R. 1666 (Doggett) would also create a cap-and-trade system, but in the early years of the program, the number of emission allowances distributed would be based on achieving a specified allowance price. The latter proposal would not allow offsets to be used for compliance purposes, while the former two would allow covered entities to satisfy an increasing percentage (approximately 30% in 2012) of their compliance obligation with offsets.

A key element in GHG emission reduction bills is how, to whom, and for what purpose the value of emission allowances or carbon tax revenue would be distributed. The distribution strategy is a critical policy decision, because it would affect (1) the overall cost of the program, and (2) how program costs are distributed throughout the economy. In the early years of the program, H.R. 2454 would distribute allowances at no cost to both covered and non-covered entities to support various policy objectives. In addition, an increasing percentage (approximately 18% in 2016) of the allowances would be sold through auction. As with the distribution of no-cost allowances, auction revenues would be used to further various policy objectives.

The Tax Extenders Act of 2009

The Tax Extenders Act of 2009 (H.R. 4213), which passed the House on December 9, 2009, would extend certain expired energy tax provisions. Specifically, the act would extend the tax credit for biodiesel tax credit and renewable diesel through the end of 2010. Additionally, the act would extend through 2010 the alternative motor vehicle credit for heavy hybrids, extend through 2010 the $0.50 per gallon production tax credit for natural gas and propane used as a transportation fuel, and extend through 2010 the deferral of gains from the sale of electric transmission property.

ENACTED LEGISLATION IN THE 110TH CONGRESS

Energy tax policy in the 110th Congress represented a shift towards increased taxes (via the removal of subsidies) on the oil and gas industry while also emphasizing energy conservation and alternative and renewable fuels, as opposed to conventional hydrocarbons.[49] This policy direction appeared to be the result of high crude oil and petroleum product prices and oil and gas industry profits, along with the political realignment of the Congress after the 2006 congressional elections. The shift was manifested by proposals to reduce oil and gas production incentives or subsidies, which were initially incorporated into, but ultimately dropped from comprehensive energy policy legislation. Later in the 110th Congress, enacted legislation focused on increasing incentives for renewable energy production, rather than reducing tax incentives available to the oil and gas industries. The fact that tax incentives for oil and gas were left in place is in part a reflection of the deteriorating business climate during 2008.

Energy Independence and Security Act of 2007 (P.L. 110-140)

The Energy Independence and Security Act of 2007 (P.L. 110-140; H.R. 6) contained a number of provisions designed to increase energy efficiency and the availability of renewable energy. Specifically, the act increased the target fuel efficiency for combined fleets of cars and light trucks, increased renewable fuel standards, and increased a number of energy-efficiency standards for household and commercial appliance equipment.

There were two provisions included in H.R. 6 that were not included in the enacted law. H.R. 6 included renewable portfolio standards that would have required a minimum amount of electricity sold by suppliers be generated using renewable resources (or credits purchased if the minimum renewable standard was not met). Additionally, H.R. 6 included provisions that would have repealed a number of subsidies that were currently available to the oil and gas industry. The approximately $22 billion saved by eliminating these subsides, under H.R. 6, would have been used to support a variety of energy efficiency and renewable energy tax incentives (such as extending the credit for electricity production from renewable resources).

Energy Tax Provisions in the Food, Conservation, and Energy Act of 2008 (P.L. 110-234)

The Food, Conservation, and Energy Act of 2008 (P.L. 110-234), otherwise referred to as the 2008 Farm Bill, contained two energy tax provisions.[50] The first provision promotes cellulosic ethanol through a blenders' credit of $1.01 per gallon, which applies to ethanol produced from qualifying cellulosic feedstocks. The second provision, the ethanol blender's tax credit (which applies to both domestic and foreign sourced ethanol), was reduced from $0.51 per gallon to $0.45 per gallon.

The Emergency Economic Stabilization Act of 2008 (P.L. 110-343)

The Emergency Economic Stabilization Act of 2008 (P.L. 110-343), included $17 billion in energy tax incentives. These provisions were primarily extensions of existing provisions (extenders), but also including several new energy tax incentives. The new provisions included $10.9 billion in renewable energy tax incentives aimed at clean energy production, $2.6 billion in incentives targeted toward cleaner vehicles and fuels, and $3.5 billion in tax breaks to promote energy conservation and energy efficiency.

The cost of the energy tax extenders legislation in the Emergency Economic Stabilization Act of 2008 was fully financed, or paid for, by raising taxes on the oil and gas industry (mostly by reducing oil and gas tax breaks) and by other tax increases. The oil and gas tax increases came from cutbacks in the IRC § 199 manufacturing deduction for income attributable to oil and gas production, which was frozen at 6% (rather than increasing to 9% as scheduled), reforming the foreign tax credit provisions, and from increasing the per-barrel tax rate on refinery crude oil under the Oil Spill Liability Trust Fund provisions.

APPENDIX. ENERGY TAX LEGISLATION PRIOR TO THE 110TH CONGRESS

This appendix briefly overviews legislation enacted during the 108th and 109th Congresses responsible for shaping current energy tax policy.

The Working Families Tax Relief Act of 2004 (P.L. 108-311)

Several energy tax incentives were extended as part of the Working Families Tax Relief Act of 2004, a $146 billion package of middle class and business tax breaks. This legislation, which was signed into law on October 4, 2004, retroactively extended four energy tax subsidies: the §45 renewable energy production tax credit, suspension of the 100% net income limitation for the oil and gas percentage depletion allowance, the $4,000 tax credit for electric vehicles, and the deduction for clean fuel vehicles (which ranges from $2,000 to $50,000). The §45 tax credit and the suspension of the 100% net income limitation had each expired on January 1, 2004 but were retroactively extended through December 31, 2005. The electric vehicle credit and the clean- vehicle income tax deduction were in the process of being phased-out (phase-out had begun on January 1, 2004). The Working Families Tax Relief Act of 2004 suspended the phase-out— providing 100% of the tax breaks—through 2005. The tax breaks were resumed beginning on January 1, 2006, when only 25% of the tax break was available.

The American Jobs Creation Act of 2004 (P.L. 108-357)

The American Jobs Creation Act of 2004 was enacted on October 22, 2004. It included about $5 billion in energy tax incentives primarily targeted at renewable energy as well as alcohol and biofuels. In particular, the act created the production tax credit, eliminated reduced tax rates for most blended alcohol fuels, established the biodiesel fuel and small refiner tax credits, and allowed a credit for oil and gas produced from marginal wells.[51]

The Energy Policy Act of 2005 (P.L. 109-58)

The Energy Policy Act of 2005 was enacted on August 8, 2005. It included an estimated $9 billion, over five years, in tax incentives distributed among renewable energy, conservation, and traditional energy sources. Among the larger provisions of the act, in revenue cost terms, were the enactment of several alternative technology vehicle credits, enactment of three investment credits for clean coal, and the extension of the production tax credit.

The Tax Increase Prevention and Reconciliation Act (P.L. 109-222)

The Tax Increase Prevention and Reconciliation Act (P.L. 109-222) was enacted May 17, 2006. It reduced the value of the subsidy by raising the amortization period from two years to five years, still faster than the capitalization treatment before the 2005 act, but slower than the treatment under that act. The higher amortization period applies only to the major integrated oil companies—independent (unintegrated) oil companies may continue to amortize all geological and geophysical (G&G) costs over two years—and it applies to abandoned as well as successful properties. This change increased taxes on major integrated oil companies by an estimated $189 million over 10 years, effectively rescinding about 20% of the nearly $1.1 billion 11-year tax for oil and gas production under the Energy Policy Act of 2005.

The Tax Relief and Health Care Act of 2006 (P.L. 109-432)

At the end of 2006, the 109th Congress enacted a tax extenders package that included extension of numerous renewable energy and excise tax provisions. Many of the renewable energy provisions in this bill had already been extended under the Energy Policy Act of 2005 and were not set to expire until the end of 2007 or later. The Tax Relief and Health Care Act of 2006 provided for one-year extensions of these provisions.

End Notes

[1] See CRS Report RL34513, *Climate Change: Current Issues and Policy Tools*, by Jane A. Leggett for an overview of climate change issues and potential policy remedies.

[2] Kenneth Gillingham, Richard G. Newell, and Karen Palmer, *Energy Efficiency Economics and Policy*, Resources for the Future, RFF DP 09-13, Washington, DC, April 2009.

[3] The extent of principal-agent problems in residential energy use is quantified in Scott Murtishaw and Jayant Sathaye, *Quantifying the Effect of the Principal-Agent Probelm on U.S. Residential Energy Use*, Lawrence Berkeley National Labratory, August 12, 2006, http://www.escholarship.org/uc/item/6f14t1 1t.

[4] Arik Levinson and Scott Niemann, "Energy Use by Apartment Tenants when Landlords Pay for Utilities," *Resource and Energy Economics*, vol. 26 (2004), pp. 5 1-75.

[5] Gilbert E. Metcalf , "Using Tax Expenditures to Achieve Energy Policy Goals," *American Economic Review*, vol. 98, no. 2 (2008), pp. 90-94.

[6] See James D. Hamilton, *Causes and Consequences of the Oil Shock of 2007-08*, National Bureau of Economic Research, Working Paper 15002, Cambridge, MA, May 2009. Hamilton evaluates the role of the oil shock of 2007-08 in the succeeding economic recession.

[7] There are non-tax options for addressing for addressing energy market failures such as regulation and private sector solutions.

[8] Taxes imposed to correct for negative externalities are also known as Pigovian taxes, named after the economist who developed the concept, Arthur Cecil Pigou.

[9] See CRS Report R40242, *Carbon Tax and Greenhouse Gas Control: Options and Considerations for Congress*, by Jonathan L. Ramseur and Larry Parker for a discussion of the relative merits and demerits of carbon taxes and cap-andtrade systems.

[10] Gilbert E. Metcalf, *Tax Policies for Low-Carbon Technologies*, National Bureau of Economic Research, Working Paper 15054, Cambridge, MA, June 2009.

[11] Gilbert E. Metcalf, "Federal Tax Policy towards Energy," *Tax Policy and the Economy*, vol. 21 (2007), pp. 145-184.

[12] It should be noted that all R&D, not just R&D related to energy, is likely to have positive externalities. There is no reason to believe that energy R&D has positive externalities that differ from R&D in general, and hence no reason to believe that energy R&D deserves a differential subsidy.

[13]See CRS Report RL3 1181, *Research and Experimentation Tax Credit: Current Status and Selected Issues for Congress*, by Gary Guenther for an overview of the research tax credit, an umbrella credit under which the energy research credit falls.

[14] Subsidizing domestic production is also problematic in that such policies conflict with environmental objectives.

[15] Gilbert E. Metcalf, "Using Tax Expenditures to Achieve Energy Policy Goals," *American Economic Review*, vol. 98, no. 2 (2008), pp. 90-94.

[16] Public goods are those that are both non-rival (one person's consumption of the good does not diminish another's ability to consume that same good) and non-excludable (it is either impossible or prohibitively expensive to prevent consumption of the good once the good has been provided). Quasi-public goods are those that are either non-rival or non-excludable.

[17] The free-rider problem is the consequence of non-excludability. If all individuals are free to use a good once that good has been provided, no single individual has an incentive to be the provider of that good. Instead, the individual will wait for the good to be provided by another party. In the absence of government intervention, the market may fail to provide goods that are subject to the free-rider problem.

[18] For background information on the federal gas tax see CRS Report R40808, *The Role of Federal Gasoline Excise Taxes in Public Policy*, by Robert Pirog.

[19] Another argument is that the federal gas tax should be viewed as correcting the externality associated with gasoline powered vehicles. Even if the gas tax were to be viewed as one correcting for the carbon emissions, it would make more economic sense to tax all carbon emissions rather than just those coming from the burning of fossil fuels by motor vehicles.

[20] Tax expenditures are government revenue losses attributable to tax provisions that allow for special exclusions, exemptions, or deductions from income or provisions that provide special tax credits, preferential tax rates, of defer tax liability.

[21] See Martin A. Sullivan, "IRS Allows New $25 Billion Tax Break for Paper Industry," Tax Notes, October 19, 2009, pp. 27 1-272 for additional information concerning the original legislative intent of the modification of the alternative fuels tax credit in SAFETEA-LU. When enacted, the modification to the alternative fuels tax credit was estimated to cost less than $100 million annually. During the first 6 months of 2009, more than $2.5 billion have been claimed for this tax credit, mostly by the paper industry.

[22] U.S. Congress, Senate Committee on Finance, "Baucus, Grassley Release Staff Draft of Legislation to Close Alternative Fuels Tax Credit Loophole," press release, June 11, 2009, http://finance.senate.gov/press /Gpress/2009/ prg061 109.pdf.

[23] Ibid.

[24] Harry De Gorter and David R. Just, "The Law of Unintended Consequences: How the U.S. Biofuel Tax Credit with a Mandate Subsidizes Oil Consumption and Has No Impact on Ethanol Consumption ," *Cornell University Working Paper Series*, 2007-20 2008.

[25] The tax expenditure for percentage depletion is computed by subtracting the value of cost depletion, the standard depletion method, from the value of percentage depletion. The resulting lifetime excess is the tax expenditure.

[26] Expensing costs means to deduct the full cost of an investment in the current tax year, rather than depreciated the costs over a period of time.

[27] Estimates in this section, and in Table 1, are based on current law as of October 31, 2008. The American Recovery and Reinvestment Act of 2009 (ARRA; P.L. 111-5) made changes to incentives for renewable energy production with an estimated additional cost of $3.6 billion between 2009 and 2013. These provisions are discussed in more detail below.

[28] Under the ARRA taxpayers may elect to take an investment tax credit in lieu of this production credit. This provision is discussed below.

[29] Electricity produced from wind, closed-loop biomass, selected solar and geothermal sources receive a credit equal to 2.1 cents per kilowatt-hour, while that sourced from open-loop biomass, small irrigation, municipal solid waste, hydro, marine, and hydrokinetic sources receive a credit of 1 cent per kilowatt-hour.

[30] For example, energy produced from a micro-turbine or geothermal deposit is eligible for a 10% credit, while fuel cell and small wind property is eligible for a credit of 30%.

[31] See CRS Report R40523, *Tax Credit Bonds: Overview and Analysis*, by Steven Maguire for additional information on tax credit bonds.

[32] Provisions enacted under ARRA added an additional $2.0 billion in tax expenditures related to energy conservation during the 2009 through 2013 budget window. These provisions are discussed below.

[33] CRS Report R40168, *Alternative Fuels and Advanced Technology Vehicles: Issues in Congress*, by Brent D. Yacobucci discusses Congressional issues specifically related to advanced technology vehicles.

[34] Modifications to tax provisions relating to alternative technology vehicles under ARRA are estimated to cost an additional $0.1 billion over the 2009 through 2013 budget window. These provisions are discussed in further detail below.

[35] Modifications to other energy provisions under ARRA are estimated to cost an additional $0.2 billion over the 2009 through 2013 budget window. These provisions are discussed in further detail below.

[36] For information on all energy provisions in ARRA see CRS Report R40412, *Energy Provisions in the American Recovery and Reinvestment Act of 2009 (P.L. 111-5)*, coordinated by Fred Sissine.

[37] The renewable energy PTC for wind facilities was extended through 2012.

[38] Qualified renewable sources include wind, open-loop biomass, closed-loop biomass, geothermal, qualified hydroelectric, landfill gas, municipal solid waste, and marine energy sources.

[39] Prior to ARRA, the credit was capped at $4,000.

[40] Prior to ARRA, the credits were capped at $2,000, $4,000 and $2,000, respectively.

[41] Prior to ARRA, the credit rate was 10%.

[42] Prior to ARRA, the credit caps ranged from $50 to $300, depending on the type of property installed. The credit was limited to $500 total for the 2006 and 2007 tax years combined. The credit was not available in 2008. The $1,500 limit applies to cumulative spending in the 2009 and 2010 tax years.

[43] See CRS Report RL3 1410, *Superfund Taxes or General Revenues: Future Funding Issues for the Sup erfund Program*, by Jonathan L. Ramseur, Mark Reisch, and James E. McCarthy and CRS Report RL3 3426, *Superfund: Implementation and Selected Issues*, by Jonathan L. Ramseur for more information on Superfund taxes.

[44] Note that the FY 2010 and FY2011 Budget Proposals represent one of many possible options to reduce the economic distortions associated with the tax treatment of energy production.

[45] Carbon tax and climate change issues are, however, largely beyond the scope of this chapter. See CRS Report R40242, *Carbon Tax and Greenhouse Gas Control: Options and Considerations for Congress*, by Jonathan L. Ramseur and Larry Parker and CRS Report R40556, *Market-Based Greenhouse Gas Control: Selected Proposals in the 111th Congress*, by Jonathan L. Ramseur, Larry Parker, and Brent D. Yacobucci for a complete discussion on carbon tax (and other market based approaches) to control greenhouse gas emissions.

[46] Greenhouse gases are those that trap heat in the earth's atmosphere. Carbon dioxide is a greenhouse gas that enters the atmosphere through the burning of fossil fuels, solid waste, and wood products, among other activities. Increases in the level of greenhouse gases in the earth's atmosphere are thought to be associated with climate change.

[47] See CRS Report R40643, *Greenhouse Gas Legislation: Summary and Analysis of H.R. 2454 as Passed by the House of Representatives* , coordinated by Mark Holt and Gene Whitney, for a detailed analysis of H.R. 2454.

[48] For a comparison of H.R. 2454 and S. 1733 see CRS Report R40896, *Climate Change: Comparison of the Cap-andTrade Provisions in H.R. 2454 and S. 1733*, by Brent D. Yacobucci, Jonathan L. Ramseur, and Larry Parker.

[49] There is an important economic distinction between a subsidy and a tax benefit. As is discussed elsewhere in this chapter, firms receive a variety of tax benefits that are not necessarily targeted subsidies (or tax expenditures) because they are available generally.

[50] See CRS Report RL34696, *The 2008 Farm Bill: Major Provisions and Legislative Action*, coordinated by Renée Johnson for a complete description of the provisions in the 2008 Farm Bill.

[51] The alcohol fuel mixture tax credit, which became law in 2005, has been the source of controversy as the credit has been claimed by a number of paper companies that burn "black liquor," a practice that was not anticipated when the legislation was drafted. When the credit was initially enacted, it was expected to cost less than $100 million annually. During the first 6 months of 2009, more than $2.5 billion has been claimed for this tax credit. For more information see Martin A. Sullivan, "IRS Allows New $25 Billion Tax Break for Paper Industry," *Tax Notes*, October 19, 2009, pp. 27 1-272 and Chuck O'Toole, "Baucus, Grassley Draft Bill to End 'Black Liquor' Subsidy," *Tax Notes*, June 15, 2009, pp. 1312-1313..

In: The Role of Taxes in Energy Production and Conservation ISBN: 978-1-61728-886-9
Editor: Bernadette M. Horton

Chapter 2

Tax Expenditures for Energy Production and Conservation*

Joint Committee on Taxation

I. Introduction and Summary Tables

The Senate Committee on Finance has scheduled a public hearing on April 23, 2009, titled "Technology Neutrality in Energy Tax: Issues and Options." Since 2004, the Congress has been very active in promulgating legislation related to energy production (including oil and gas and renewables) and conservation. Part II of this document,[1] prepared by the staff of the Joint Committee on Taxation, provides a description of present law tax expenditures for energy production and conservation. These tax provisions vary widely, and are summarized in the tables that follow. In addition to the energy specific tax expenditures, energy sector producers and manufacturers may also benefit from other general tax expenditures, such as the section 199 deduction for income attributable to domestic production activities, additional first year depreciation ("bonus depreciation"), the lower rates of tax on the first $10 million of corporate taxable income, and, in general, the deferral of tax on active business income derived by foreign subsidiaries.

The various tax benefits create incentives that have the potential to affect economic decisions and allocate economic resources from other uses to the tax-favored uses. Such tax preferences may produce an allocation of resources that is more efficient for society at large if they are properly designed to overcome negative effects (such as atmospheric pollution, for example) that would otherwise result from a purely market based outcome without any government intervention. The extensive variety of tax expenditures for energy production and conservation have been criticized for lacking well defined objectives, and for lacking coordination among provisions having similar objectives. Some argue that the simultaneous existence of tax preferences for the fossil fuel industry and for renewable energy production represents an incoherent government policy. Others have noted that the incentives for

* This is an edited, reformatted and augmented edition of a United States Senate Committee on Finance publication, Report JCX-25-09R, dated April 21, 2009.

renewable energy and conservation are not themselves designed in a coordinated way to produce the most efficient or equitable subsidies for renewable energy and conservation.

Part III of this document provides the staff of the Joint Committee on Taxation's most recent estimates of tax expenditures in the energy sector, a brief discussion of the economic rationale for certain government intervention in energy markets through the tax code, and issues related to the proper design of such tax preferences.

A. Summary of Credit for Electricity Produced from Certain Renewable Resources

Eligible electricity production activity (sec. 45)	Credit amount for 2008[3] (cents per kilowatt-hour)	Credit period for facilities placed in service on or before August 8, 2005 (years from placed-in-service date)	Credit period for facilities placed in service after August 8, 2005 (years from placed-in-service date)
Wind	2.1	10	10
Closed-loop biomass	2.1	10[1]	10
Open-loop biomass (including agricultural livestock waste nutrient facilities)	1.0	5[2]	10
Geothermal	2.1	5	10
Solar (pre-2006 facilities only)	2.1	5	10
Small irrigation power	1.0	5	10
Municipal solid waste (including landfill gas facilities and trash combustion facilities)	1.0	5	10
Qualified hydropower	1.0	N/A	10
Marine and hydrokinetic	1.0	N/A	10

[1] In the case of certain co-firing closed-loop facilities, the credit period begins no earlier than October 22, 2004.

[2] For certain facilities placed in service before October 22, 2004, the five-year credit period commences on January 1, 2005.

[3] Inflation adjusted amounts for 2009 are not yet available.

B. Summary of Certain Renewable and Alternative Fuel Incentives

Fuel Type	Per Gallon Incentive Amount	Incentive Expires
Agri-biodiesel and biodiesel	$1.00 per gallon, plus $0.10 per gallon for small agri-biodiesel producers	December 31, 2009
Renewable diesel	$1.00 per gallon	December 31, 2009
Alcohol fuel (not ethanol and other than from natural gas or coal)	$0.60 per gallon	December 31, 2010
Ethanol fuel	$0.45 per gallon, plus $0.10 per gallon for small producers	December 31, 2010

Table B. (Continued)

Fuel Type	Per Gallon Incentive Amount	Incentive Expires
Cellulosic biofuel	$1.01 per gallon (for alcohol, $1.01 per gallon less the amount of the alcohol fuel mixture credit and small ethanol producer's credit in effect at the time of production)	December 31, 2012
Alternative fuel: • liquefied petroleum gas • P Series Fuels • compressed or liquefied natural gas • liquefied hydrogen • any liquid fuel derived from coal through the Fischer-Tropsch process • compressed or liquified gas derived from biomass • liquid fuel derived from biomass	$0.50 per gallon	December 31, 2009 (September 30, 2014, in the case of liquefied hydrogen)

C. Summary of Investment Tax Credit Energy Production Incentives

		Credit rate	Maximum credit	Expiration
Energy credit (sec. 48)	Equipment to produce a geothermal deposit	10%	none	None
	Equipment to use ground or ground water for heating or cooling	10%	none	December 31, 2016
	Microturbine property (< 2 Mw electrical generation power plants of >26% efficiency)	10%	$200 per Kw of capacity	December 31, 2016
	Combined heat and power property (simultaneous production of electrical/mechanical power and useful heat > 60% efficiency)	10%	none	December 31, 2016
	Solar electric or solar hot water property	30% (10% after December 31, 2016)	none	None
	Fuel cell property (generates electricity through electrochemical process)	30%	$1,500 for each 1/2 Kw of capacity	December 31, 2016
	Small (<100 Kw capacity) wind electrical generation property	30%	none	December 31, 2016

D. Summary of Energy Conservation Credits

		Credit rate or amount	Maximum credit	Expiration
Personal credits:				
Nonbusiness energy property credits (sec. 25C)	Insulation to inter-national energy conservation code standard	30 %	$1,500	December 31, 2010
	Energy efficient Windows, doors, skylights, roofs	30 %	$1,500	December 31, 2010
	Advanced main air circulating fans	30 %	$1,500	December 31, 2010
	Qualified natural gas, propane, or oil furnace or hot water boilers	30 %	$1,500	December 31, 2010
	Electric heat pump water heaters or natural gas, propane, or oil water heaters	30 %	$1,500	December 31, 2010
	Central air conditioners	30 %	$1,500	December 31, 2010
	Biomass fuel property (wood stoves)	30 %	$1,500	
Residential energy efficient property credits (sec. 25D)	Residential solar water heating or solar electric property	30 %	none	December 31, 2016
	Residential small wind property	30 %	none	December 31, 2016
	Residential geothermal heat pump property	30 %	none	December 31, 2016
	Residential fuel cell property	30 %	$500 per half kilowatt of capacity	December 31, 2016
Business Credits:				
Manufacturer credit for new energy effici-ent home (sec. 45L)	Homes 30% more efficient than standard	$1,000 per home	none	December 31, 2009
	Homes 50% more efficient than standard	$2,000 per home	none	December 31, 2009
Manufacturer credit for energy efficient appliances (sec. 45M)	Dishwashers	$45	(1)	December 31, 2009
	Dishwashers (higher efficiency standard)	$75	(1)	December 31, 2010
	Clothes washers	$125	(1)	December 31, 2009
	Clothes washers (higher efficiency standard)	$150	(1)	December 31, 2010
	Clothes washers (highest efficiency standard)	$250	none	December 31, 2010
	Refrigerators	$75	(1)	December 31, 2009
	Refrigerators (higher efficiency standard)	$100	(1)	December 31, 2010
	Refrigerators (highest efficiency standard)	$200	none	December 31, 2010

[1] A given manufacturer may not claim credits in excess of an aggregate of $75 million for taxable years beginning after December 31, 2007, with respect to all credits excepting the $200 credit for refrigerators and the $250 credit for clothes washers.

E. Summary of Alternative Fuel Vehicle Credits

Type of Property	Description of Qualifying Property	Credit Amount and Explanation	Expiration
Fuel cell vehicles (sec. 30B)	Vehicles propelled by chemically combining oxygen with hydrogen and creating electricity	• Base credit of $8,000 (reduced to $4,000 after 2009) for vehicles weighing 8,500 pounds or less • Heavier vehicles can get up to a $40,000 credit, depending on weight • An additional $1,000 to $4,000 credit is available to cars and light trucks to the extent fuel economy exceeds 2002 base fuel economy	12/31/2014
Hybrid vehicles (sec. 30B)	Vehicles powered by internal combustion engine and a rechargeable energy storage system (e.g. batteries)	• Base credit of $400 to $2,400 for cars and light trucks (weighing 8,500 pounds or less) depending on amount by which fuel economy exceeds 2002 base fuel economy • Additional conservation credit for cars and light trucks of between $250 and $1,000 depending on estimated lifetime fuel savings relative to a comparable 2002 model year vehicle powered solely by a gasoline or diesel engine • For heavy hybrid vehicles, the credit varies between 20 and 40 percent of the incremental cost depending on the amount by which the fuel economy of the vehicle exceeds that of a comparable vehicle. The maximum credit is $7,500 for vehicles weighing 14,000 pounds of less; $15,000 for vehicles weighing more than 14,000 pounds but less than 26,000 pounds; $30,000 for vehicles weighing more than 26,000 pounds.	• 12/31/2009 for cars and light trucks weighing 8,500 pounds or less • 12/31/2010 for heavier vehicles • 60,000 vehicle per manufacturer limitation applies to cars and light trucks
Lean burn vehicles (sec. 30B)	• Cars and light trucks with internal combustion engines designed to use more air than necessary for complete co-mbustion that incorporate direct injection and achieve at least 125	Credit calculation is the same as for hybrid passenger vehicles and light trucks.	• 12/31/2010 • 60,000 vehicle per manufacturer limitation

Table E. (Continued)

Type of Property	Description of Qualifying Property	Credit Amount and Explanation	Expiration
	percent of the 2002 model year city fuel economy • Commonly referred to as clean diesel cars and trucks		
Alternative fuel vehi-cles (sec. 30B)	• Vehicles that run solely on compr-essed natural gas, liquefied natural gas, liquefied petroleum gas, hydrogen, or any liquid fuel that is at least 85 percent methanol	• Credit is 50 percent of the incremental cost of such vehicle above a comparable vehicle, plus an additional 30 percent if the vehicle meets certain emissions standards • Depending on vehicle weight, the maximum allowable credit varies between $5,000 (for vehicles weighing 8,500	12/31/2010
	• Vehicles that run on a fuel mixture 75 percent of which is one of the fuels listed above.	pounds or less) to $40,000 (for vehicles weighing over 26,000 pounds) • Mixed-fuel vehicles using 90 percent alternative fuel get 90 percent of the credit; those using 75 percent alte-rnative fuel get 70 percent of the credit	
Plug-in electric-drive motor vehicles (2009) (sec. 30D)	A four-wheeled vehicle propelled by a battery with at least 4 kilowatt-hours of electricity that can be charged from an external source	• Base credit of $2,500 plus $417 for each kilowatt-hour of additional battery capacity in excess of 4 kilowatt-hours. • Depending on vehicle weight, the maximum allowable credit varies between $7,500 (for vehicles weighing 10,000 pounds or less) to $15,000 (for vehicles weighing over 26,000 pounds)	• 250,000 vehicle overall cap • Modified as de-scribed below after 2009
Plug-in ele-ctric- drive motor vehi-cles (after 2009) (sec. 30D)	As described above, but excludes low speed vehicles and vehicles weighing 14,000 or more	Credit is as described above except that the maximum credit is $7,500 regardless of vehicle weight.	200,000 vehicle per manufacturer limitation
Electric-drive low-speed, motorcycle, and three-wheeled vehicles (sec. 30)	• Vehicles otherw-ise qualifying as plug-in electric-drive vehicles but for the fact that they have limited speed or less than four wheels • Two- and three-wheeled vehicles must have a battery capacity of at least 2.5 kilowatt- hours	Credit is 10 percent of cost, up to $2,500.	12/31/2011

Table E. (Continued)

Type of Property	Description of Qualifying Property	Credit Amount and Explanation	Expiration
Converted plug-in electric-drive vehicles (sec. 30B)	Used vehicles that have been converted into a plug-in electric drive motor vehicle	Credit is 10 percent of conversion cost up to $4,000.	12/31/2011
Alternative fuel refueling property (sec. 30C)	Property that dispenses alternative fuels, including ethanol, biodiesel, natural gas, hydrogen, and electricity	• For 2009 and 2010, credit is 50 percent of cost of qualifyed refueling property (other than hydrogen property) inst-alled at a business or principal residence; hydrogen pro-perty gets a 30 percent credit • For 2009 and 2010, credit may not exceed $200,000 for hydrogen refueling property used in a business, $50,000 for other business refueling property, and $2,000 for non-hydrogen refueling property installed at a principal residence • For years other than 2009 and 2010, the credit rate is 30 percent, and the maximum allowable credit is $50,000 for business refueling property and $1,000 for property installed at a principal residence.	• 12/31/2010 for non-hydrogen refueling property • 12/31/2014 for hydrogen refueling property

F. Summary of Certain Non-Fossil Fuel Capital Cost Recovery Provisions

	Description of Provision	Expiration
Five-year cost recovery for certain energy property (sec. 168(e)(3)(B)(vi))	• A five-year MACRS recovery period is generally provided for equipment using solar and wind energy to generate electricity, to heat or cool (or provide hot water for use in) a structure, or to provide solar process heat; equipment using solar energy to illuminate the inside of a structure using fiber-optic distributed sunlight; equipment used to produce, distribute, or use energy derived from a geothermal deposit; and qualified fuel cell property. • A five-year MACRS recovery period is provi-ded for certain biomass property, including (i) a boiler, the primary fuel for which will be an alternate substance; (ii) a burner (including necessary on-site equipment to bring the alternate substance to the burner) for a combustor other than a boiler if the primary fuel for such burner will be an alternate substance; (iii) equipment for converting an alternate substance into a qualified fuel; and (iv) certain pollution control equipment.	For five-year recovery period for certain solar equipment - December 31, 2016

Table F. (Continued)

	Description of Provision	Expiration
Special allowance for cellulosic biofuel plant property (sec. 168(l))	An additional first-year depreciation deduction equal to 50 percent of the adjusted basis of qualified cellulosic biofuel plant property.	December 31, 2012
Pollution control facilities (secs. 169, 291)	A taxpayer may elect to recover the cost of any certified pollution control facility over a period of 60 months. A corporation taxpayer must reduce the amount of basis otherwise eligible for the 60- month recovery by 20 percent.	None
Energy efficient commercial buildings dedu-ction (sec. 179D)	A taxpayer may take an additional deduction of $1.80 per square foot of commercial property that exceeds certain energy efficiency standards	December 31, 2013

G. Summary of Fossil Fuel Capital Cost Recovery Provisions

	Description of Provision	Expiration
Geological & geophysical expenditures (sec. 167(h))	• Geological and geophysical (G&G) expenditures incurred by independent producers and smaller integrated oil companies in connection with domestic oil and gas exploration may be amor-tized over 24 months. • G&G expenditures incurred by major integrated oil companies are amortized over seven years.	None
Alaska natural gas pipeline (sec. 168(e)(3)(C)(iii))	Seven-year MACRS recovery period and a class life of 22 years is provided for any natural gas pipeline system located in the State of Alaska that has a capacity of more than 500 billion Btu of natural gas per day and is placed in service after December 31, 2013.	None
Natural gas dis-tribution lines (sec. 168(e)(3) (E)(viii))	Fifteen-year MACRS recovery period and a class life of 35 years is provided for natural gas distri-bution lines placed in service after April 11, 2005 and before January 1, 2011 provided original use of the property begins with the taxpayer	December 31, 2010
Natural gas gathering lines (sec. 68(e)(3) (C)(iv))	Seven-year MACRS recovery period and 14-year class life is provided for natural gas gathering pipelines.	None
Election to expense 50 percent of qualified property used in refining liquid fuels (sec. 179C)	Taxpayers may elect to expense 50 percent of the cost of qualified refinery property used for processing liquid fuel from crude oil or qualified fuels; remaining 50 percent is recovered under otherwise applicable rules.	December 31, 2013
Special expensing rule for capital costs incurred by small refiners to comply with EPA sulfur regulations (sec. 179B)	A small business refiner may deduct 75 percent of the costs paid or incurred in complying with the Highway Diesel Fuel Sulfur Control requirement of the EPA.	Costs must be paid or incurred on or before December 31, 2009

Table G. (Continued)

	Description of Provision	Expiration
Deduction for tertiary inject-tants (sec. 193)	Taxpayers engaged in petroleum extraction activities may generally deduct qualified tertiary injectant expenses used while applying a tertiary recovery method.	None
Election to expe-nse intangible drilling costs (secs. 263(c) and 291)	Taxpayers may elect to currently deduct intangible drilling costs (IDCs) paid or incurred with respect to the development of an oil or gas property located in the United States. For an integrated oil company that has elected to expense IDCs, 30 percent of the IDCs on productive wells must be capitalized and amortized over a 60-month period.	None
Depletion (secs. 611-613A)	• Depletion is available to any person having an economic interest in a producing oil and gas property. There are generally two types of depletion--cost and percentage depletion. Cost depletion is limited to the taxpayer's basis in the property, whereas percentage depletion is not limited by the basis, but is subject to limitations based on net income derived from the property and taxable income. • Percentage depletion for producing oil and gas property (15 percent rate) is available only to independent producers and royalty owners. For marginal properties, the taxable income limitation is suspended for taxable years ending before January 1, 2010. • Percentage depletion is also available for coal and lignite (10 percent rate) and oil shale (15 percent rate). The percentage depletion deduction coal and lignite is generally reduced by an amount equal to 20 percent of the percentage depletion that exceeds the adjusted basis of the property.	Suspension of taxable income limitation for marginal properties expires December 31, 2009

H. Summary of Energy Credits Related to Fossil Fuels

Eligible Activity	Description	Credit Amount	Expiration
Enhanced oil recovery (EOR) cre-dit (sec. 43)	• Credit for expenses associated with an EOR project • An EOR project is generally a project that involves the use of one or more tertiary recovery methods to increase the amount of recoverable domestic crude oil	• 15 percent of enhanced oil recovery costs • Currently phased-out	None
Marginal wells credit (sec. 45I)	Production credit for marginal wells or wells that have an average daily production of not more than 25 barrels per day	• $3-per-barrel credit (adjusted for inflation from 2004) for the production of crude oil from marginal wells • $0.50-per-1,000-cubic-feet credit (adjusted for inflation from 2004) for the production of	None

Table H. (Continued)

Eligible Activity	Description	Credit Amount	Expiration
		natural gas from a marginal wells Currently phased-out	
Indian coal credit (sec. 45)	Production credit for coal produced at facilities placed in service before 2009 that produce coal from reserves that on June 14, 2005 were owned by (or held in trust on behalf of) an Indian tribe	• $1.50-per-ton production credit (adjusted for inflation from 2005) for 2006 and 2007 • $2-per-ton credit (adjust for inflation from 2005) for 2008 through 2012	12/31/12
Refined coal credit (used to produce steam) (sec. 45)	Production credit for refined coal, defined as a synthetic fuel produced from coal (including lignite) or high-carbon fly ash that when burned emits 20 percent less nitrogen oxide and 40 percent less sulfur dioxide or mercury compared to feedstock coal available in 2003	• $4.375-per-ton production credit (adjusted for inflation from 1992) • Credit is available during the 10-year period from the date the facility was placed in service	12/31/2009
Refined coal credit (used as steel ind-ustry fuel) (sec. 45)	Production credit for steel industry fuel, defined as a fuel produced through a process of liquefying coal waste sludge, distributing the liquefied product on coal, and using the resulting mixture as a feedstock for the manufacture of coke	$2 (adjusted for inflation from 1992) per barrel-of-oil equivalent	• Generally, one year from facility placed-in-service date or 12/31/2009 • Facilities must be pla-ced in service before 1/1/2010
Coke credit (sec. 45K)	• Credit for the production of coke or coke gas (excluding petroleum-based coke or coke gas) • Amount of credit-eligible coke per facility is limited to 4,000 barrels per day	$3 (adjusted for inflation from 2004) per barrel-of-oil equivalent	• Facilities must be placed in service by 12/31/2009 • Credit ava-ilable for four years from placed-in-service date
Advanced coal project credit (sec. 48A)	• Investment credit for projects that use integrated gasification combined cycle (IGCC) or other advanced coal-based electricity generation technologies • Credits are allocated by the Secretary • First round allocations are capped at $800 million for IGCC projects and $500 million for other projects • Second round allocations are capped at $1.25 billion	• 20 percent for first round IGCC projects • 15 percent for other first round projects • 30 percent for second round projects	None (other than the credit allocation limitation)

Table H. (Continued)

Eligible Activity	Description	Credit Amount	Expiration
	• Second round projects must generally sequester 65 percent of total CO_2 emissions		
Gasification credit (sec. 48B)	• Investment credit for qualified projects that use gasification technology • Qualified projects convert coal, petroleum residue, biomass, or other materials recovered for their energy content into a synthesis gas for direct use or subsequent chemical or physical conversion • Credits are allocated by the Secretary • First round allocations are capped at $350 million • Second round allocations are capped at $250 million • First round projects are generally limited to industrial applications; second round projects include projects designed to produce motor fuels • Second round projects must generally sequester 65 percent of total CO_2 emissions	• 20 percent for first round • 30 percent for second round	None (other than the credit allocation limitation)

I. Summary of Other Energy Provisions

Eligible Activity	Description	Credit Amount for 2008	Expiration
Carbon dioxide sequestration credit (sec. 45Q)	• Credit for the sequestration of industrial source carbon dioxide produced at qualified U.S. facilities • Qualified facilities must capture at least 500,000 metric tons of CO_2 per year.	• $10 for CO_2 used as a tertiary injectant and then permane-ntly sequestered • $20 for CO_2 permanently sequestered without being first used as a tertiary injectant	End of the year in which the Secretary determi-nes that 75 mill-ion tons of CO_2 have been captured and sequestered
Energy research credit (sec. 41)	• Credit for payments made to energy research consortia for qualified energy research • Includes research related to fossil fuels as well as to renewable energy technologies	20 percent of qualified expenses	12/31/2009
Advanced energy property credit (sec. 48C)	• Investment credit for property used in a qualified advanced energy manufacturing project	30 percent of qua-lified investment expenses	None other than the credit allo-cation limitation

Table I. (Continued)

Eligible Activity	Description	Credit Amount for 2008	Expiration
	• Credits are allocated by the Secretary • Up to $2.3 billion of credits may be allocated		
Advanced nuclear power production credit (sec. 45J)	• Credit for production of nuclear power from new facilities that use modern designs and have recei-ved an allocation from the Secretary • Secretary may allocate up 6,000 megawatts of credit-eligible capacity	1.8 cents per kilowatt-hour for the eight-year peri-od starting when the facility was placed in service	Qualified facil-ities must have been placed in service by 12/31/2020
Passive loss rules for working interests in oil and gas property (sec. 469)	• Passive activity loss rules not applicable to working interest in any oil or gas property that taxpayer holds directly or indirectly through an entity that does not limit the taxpayer's liability • Losses and credits from such interests, in general, may offset income from other activities of taxpayer	N/A	None
Nuclear decommissioning costs (sec. 468A)	• Current tax deduction allowed for contributions to qualified nuclear decommissioning fund at election of taxpayer • Income of fund taxed at reduced rate of 20 percent	N/A	None
Reduced tax for diesel-water fuel emulsion (secs. 4081(a)(2)(D), 4081(c), and 6427(m))	• Diesel fuel tax rate of 24.3 cents per gallon is reduced to 19.7 cents per gallon for diesel- water fuel emulsion to reflect the reduced Btu content per gallon resulting from the water. • Refund of the difference between the two rates is available to the extent tax-paid diesel is used to produce a qualifying emulsion diesel fuel.	N/A	None
Certain publicly treated partn-erships treated as corporations (secs. 7704 and 851)	• General rule that a publ-icly traded partnership is taxed as a corporation is not applicable if 90 percent of gross income is interest, dividends, real property rents, or certain other types of qualifying income	N/A	None

Table I. (Continued)

Eligible Activity	Description	Credit Amount for 2008	Expiration
	• Other types of qualifying income includes income and gains from certain activities with respect to natural resources		
Energy conservation subsidies provided by public utilities (sec. 136)	• Energy conservation subsidies provided by public utilities are excluded from gross income	N/A	None
Five-Year carry-back of net operating losses for certain electric utility companies (sec. 72(b)(1)(I))	• A portion of net operating losses arising in 2003 - 2005 for certain electric utility companies may be carried back for 5 years	N/A	December 31, 2008
Deferral of Gains from the Sale of Electric Transmission Property (sec.451(i))	• A taxpayer may elect to recognize gain ratably over an eight year period for gains on disposition of certain electric transmission property	N/A	December 31, 2009

J. Summary of Energy-Related Bond Provisions

Type of Bond	Description
Clean Renewable Energy Bonds ("CREBs") (sec. 54)	• Tax credit bond. • CREBs are defined as any bond if the proceeds from the issuance of such bonds are used to finance facilities that qualify for the tax credit under section 45 (other than Indian coal production facilities). • Credit rate is rate that permits issuance of CREBs without discount and interest cost to the qualified issuer. • Qualified issuers include electrical cooperatives, clean renewable energy bond lenders, and State and local governments (including Indian tribes). • Volume limited and allocated by the Secretary of the Treasury.
New CREBs (sec. 54C)	• Tax credit bond. • New CREBs may be issued to finance "qualified renewable energy facilities." Qualified renewable energy facilities are facilities that: (1) qualify for the tax credit under section 45 (other than Indian coal and refined coal production facilities). • Credit rate is 70 percent of the rate that permits issuance of bonds without discount and interest cost to the issuer • Qualified issuers include electrical cooperatives, clean renewable energy bond lenders, public power providers, State and local governments (including Indian tribes), and not-for-profit electric utilities which have a loan or loan guarantee under the Rural Electrification Act. • Volume limited and allocated by the Secretary of the Treasury.
Qualified Energy Conservation Tax Credit Bonds ("QECs")(sec. 54D)	• Tax credit bond. • Bond issuance must be used for "qualified conservation purposes" (described in detail in section II.B.6 of this document).

Table J. (Continued)

Type of Bond	Description
	• Credit rate is 70 percent of the rate that permits issuance of bonds without discount and interest cost to the issuer. • Volume limited and allocated by the Secretary of the Treasury.
Safe Harbor from arbitrage rules for prepaid natural gas (sec. 148)	• Allows tax-exempt bonds to be used to finance prepaid natural gas contracts without application of the otherwise applicable arbitrage rules.
Tax-exempt bonds for certain public energy - related projects (sec. 103)	• Tax-exempt bond • May be used for financing government-owned and operated electrical and gas powered generation, transmission and distribution facilities • Not subject to any volume caps
Tax-exempt bonds for certain public private energy - related projects (secs. 141, and 142)	• Tax-exempt bond • May be used for financing certain exempt facilities including privately owned and/or operated utility facilities (sewage, water, solid waste disposal, and local district heating and cooling facilities, certain private electric and gas facilities, and hydroelectric dam enhancements); qualified green building and sustainable design projects • Generally subject to private activity volume cap

II. Description of Present Law

A. Tax Provisions Relating to Renewable Energy

1. Credit for alternative fuel vehicle refueling property (sec. 30C)

Present Law

Taxpayers may claim a 30-percent credit for the cost of installing qualified clean-fuel vehicle refueling property to be used in a trade or business of the taxpayer or installed at the principal residence of the taxpayer.[2] The credit may not exceed $30,000 per taxable year per location, in the case of qualified refueling property used in a trade or business and $1,000 per taxable year per location, in the case of qualified refueling property installed on property which is used as a principal residence.

For property placed in service in 2009 or 2010, the maximum credit available for business property is increased to $200,000 for qualified hydrogen refueling property and to $50,000 for other qualified refueling property. For nonbusiness property, the maximum credit is increased to $2,000 for refueling property other than hydrogen refueling property. In addition, during these years, the credit rate is increased from 30 percent to 50 percent for refueling property other than hydrogen refueling property.

Qualified refueling property is property (not including a building or its structural components) for the storage or dispensing of a clean-burning fuel or electricity into the fuel tank or battery of a motor vehicle propelled by such fuel or electricity, but only if the storage or dispensing of the fuel or electricity is at the point of delivery into the fuel tank or battery of the motor vehicle. The use of such property must begin with the taxpayer.

Clean-burning fuels are any fuel at least 85 percent of the volume of which consists of ethanol, natural gas, compressed natural gas, liquefied natural gas, liquefied petroleum gas, or hydrogen. In addition, any mixture of biodiesel and diesel fuel, determined without regard to any use of kerosene and containing at least 20 percent biodiesel, qualifies as a clean fuel.

Credits for qualified refueling property used in a trade or business are part of the general business credit and may be carried back for one year and forward for 20 years. Credits for residential qualified refueling property cannot exceed for any taxable year the difference between the taxpayer's regular tax (reduced by certain other credits) and the taxpayer's tentative minimum tax. Generally, in the case of qualified refueling property sold to a tax-exempt entity, the taxpayer selling the property may claim the credit.

A taxpayer's basis in qualified refueling property is reduced by the amount of the credit. In addition, no credit is available for property used outside the United States or for which an election to expense has been made under section 179.

The credit is available for property placed in service after December 31, 2005, and (except in the case of hydrogen refueling property) before January 1, 2011. In the case of hydrogen refueling property, the property must be placed in service before January 1, 2015.

2. Incentives for alcohol, cellulosic biofuel, biodiesel, renewable diesel, and certain alternative fuels (secs. 40, 40A, 6426, and 6427(e))

Overview

The Code provides per-gallon incentives relating to the following qualified fuels: alcohol (including ethanol), biodiesel (including agri-biodiesel), renewable diesel, and certain alternative fuels.[3] The incentives may be taken as an income tax credit, excise tax credit or payment. The provisions are coordinated so that a gallon of qualified fuel is only taken into account once. If the qualified fuel is part of a qualified fuel mixture, the incentives apply only to the amount of qualified fuel in the mixture. The Code also provides an income tax credit for qualified cellulosic biofuel production.

For alcohol, other than ethanol, the amount of the credit generally is 60 cents per gallon.[4] For ethanol, the credit is generally 45 cents per gallon, with an extra 10 cents per gallon available for small ethanol producers. The alcohol incentives expire after December 31, 2010. The amount of the credit is $1.00 per gallon for biodiesel, agri-biodiesel and renewable diesel. An extra 10 cents per gallon is available for small producers of agri-biodiesel. The biodiesel, agribiodiesel and renewable diesel incentives expire after December 31, 2009. The credit amount for alternative fuels is 50 cents per gallon. The incentives for alternative fuels expire after December 31, 2009 (after September 30, 2014, in the case of liquefied hydrogen). For cellulosic biofuel, the provision is applicable to qualified cellulosic biofuel production through December 31, 2012.

Fuel that is produced outside the United States for use as a fuel outside the United States is ineligible for the per-gallon tax incentives relating to alcohol, biodiesel, renewable diesel, and alternative fuel. For example, fuel in the following situations is ineligible for incentives: (1) biodiesel, which is not in a mixture, that is both produced and used outside the United States, (2) foreign-produced biodiesel that is used to make a qualified mixture outside of the United States for foreign use, and (3) foreign-produced biodiesel that is used to make a qualified mixture in the United States that is then exported for foreign use. For cellulosic biofuel, the fuel must be both produced in the United States and used as fuel in the United States.

Alcohol made from petroleum, natural gas, or coal does not qualify for the incentives mentioned above. For fuel alcohol that is made from natural gas, the Code provides a reduced rate of tax. This incentive expires September 30, 2011. The reduced rate of tax for alcohol made from coal expired January 1, 2009.

Alcohol and cellulosic biofuel

Sections 40, 6426 and 6427(e) provide per-gallon tax incentives for the sale, use and production of alcohol fuel and alcohol fuel mixtures. The incentives for alcohol generally do not apply after December 31, 2010. For cellulosic biofuel (discussed infra), the incentive is unavailable after December 31, 2012.

"Alcohol" includes methanol and ethanol, and the alcohol gallon equivalent of ethyl tertiary butyl ether, or other ethers produced from such alcohol. It does not include alcohol produced from petroleum, natural gas, or coal, or any alcohol with a proof of less than 150 (190 proof for purposes of the credit taken under 6426 or payment under section 6427). Denaturants (additives that make the alcohol unfit for human consumption) are disregarded for purposes of determining proof. However, denaturants are taken into account in determining the volume of alcohol eligible for the per-gallon incentive. In calculating alcohol volume, denaturants cannot exceed two percent of volume.[5]

The section 40 alcohol fuels credit is an income tax credit comprised of four components: (1) the alcohol mixture credit, (2) the alcohol credit, (3) the small ethanol producer credit, and (4) the cellulosic biofuel producer credit. Sections 6426 and 6427(e) pertain to alcohol fuel mixtures only.

Alcohol mixture credits and payments

The alcohol fuel mixture credit may be taken as part of the section 40 income tax credit, the section 6426 excise tax credit or as a payment under section 6427. For section 40, an alcohol fuel mixture is a mixture of alcohol and gasoline or alcohol and a special fuel. Since the excise tax credit is taken against the liability for taxable fuels (gasoline, kerosene, or diesel), for purposes of the excise tax payments and credits, an alcohol fuel mixture is a mixture of alcohol and a taxable fuel.

The fuel must be either sold for use as a fuel to another person or used as fuel in the mixture producer's trade or business. The addition of denaturants is not the production of a mixture. The credit is allowed only for the gallons of alcohol used to produce the mixture. For alcohol that is ethanol, the amount of the incentive is 45 cents per gallon. For other alcohol, the incentive is 60 cents per gallon.

The alcohol mixture credit is most often taken as an excise tax credit or payment. Persons who blend alcohol with gasoline, diesel fuel, or kerosene to produce an alcohol fuel mixture must pay tax on the volume of alcohol in the mixture when the mixture is sold or removed. The alcohol fuel mixture credit must first be taken to reduce excise tax liability for gasoline, diesel fuel or kerosene. Any excess credit may be taken as a payment or income tax credit.

Alcohol credit (straight or "neat" alcohol)

The second component of the section 40 income tax credit is the alcohol credit. The credit is available for alcohol (not in a mixture) that is either (1) used as a fuel in the taxpayer's trade or business, or (2) sold at retail and placed in the fuel tank of the retail buyer.

The credit cannot be claimed for alcohol bought at retail, even if the buyer uses it as a fuel in a trade or business. This credit is not available as an excise tax credit or payment.

Small ethanol producer credit

The third component of the section 40 income tax credit is the small ethanol producer credit. It is in addition to the credits described above and is an extra 10 cents per gallon available for up to 15 million gallons of qualified ethanol fuel production for any tax year. The 15 million gallon limitation is waived for ethanol that is cellulosic ethanol. The credit is available to eligible small ethanol producers, defined as producers who have an annual productive capacity of not more than 60 million gallons of any type of alcohol. Qualified ethanol fuel production is ethanol produced and sold by such producer to another person (a) for use by such other person in the production of a qualified alcohol fuel mixture in such person's trade or business (other than casual off-farm production), (b) for use by such other person as a fuel in a trade or business, or (c) who sells such ethanol at retail to another person and places such ethanol in the fuel tank of such other person. Qualified ethanol fuel production also includes use or sale by the producer for any purpose described in (a), (b), or (c). A cooperative may pass through the small ethanol producer credit to its patrons.

The small ethanol producer credit is not available as an excise tax credit or payment.

Cellulosic biofuel producer credit

Section 15332 of the Food, Conservation, and Energy Act of 2008 (Pub. L. No 110-234) added a new component to section 40 of the Code, the "cellulosic biofuel producer credit." This credit is a nonrefundable income tax credit for each gallon of qualified cellulosic fuel production of the producer for the taxable year. The amount of the credit per gallon is $1.01, except in the case of cellulosic biofuel that is alcohol. In the case of cellulosic biofuel that is alcohol, the $1.01 credit amount is reduced by (1) the credit amount applicable for such alcohol under the alcohol mixture credit as in effect at the time cellulosic biofuel is produced and (2) in the case of cellulosic biofuel that is also ethanol, the credit amount for small ethanol producers as in effect at the time the cellulosic biofuel fuel is produced. The reduction applies regardless of whether the producer claims the alcohol mixture credit or small ethanol producer credit with respect to the cellulosic alcohol. When the alcohol mixture credit and small ethanol producer credit expire after December 31, 2010, cellulosic biofuel will receive the $1.01 without reduction.

"Qualified cellulosic biofuel production" is any cellulosic biofuel which is produced by the taxpayer and which is (1) sold by the taxpayer to another person (a) for use by such other person in the production of a qualified biofuel fuel mixture in such person's trade or business (other than casual off-farm production), (b) for use by such other person as a fuel in a trade or business, or, (c) who sells such biofuel at retail to another person and places such biofuel in the fuel tank of such other person; or (2) used by the producer for any purpose described in (a), (b), or (c).

"Cellulosic biofuel" means any liquid fuel that (1) is produced in the United States and used as fuel in the United States,[6] (2) is derived from any lignocellulosic or hemicellulosic matter that is available on a renewable or recurring basis and (3) meets the registration requirements for fuels and fuel additives established by the Environmental Protection Agency ("EPA") under section 211 of the Clean Air Act. Thus, to qualify for the credit the fuel must be approved by the Environmental Protection Agency. Cellulosic biofuel does not include any alcohol with a proof of less than 150.

A "qualified cellulosic biofuel mixture" is a mixture of cellulosic biofuel and a special fuel or of cellulosic biofuel and gasoline, which is sold by the person producing such mixture to any person for use as a fuel, or is used as a fuel by the person producing such mixture. The term "special fuel" includes any liquid fuel (other than gasoline) which is suitable for use in an internal combustion engine.

The cellulosic biofuel producer credit terminates on December 31, 2012. The cellulosic biofuel producer credit cannot be claimed unless the taxpayer is registered with the IRS as a producer of cellulosic biofuel.

Cellulosic biofuel eligible for the section 40 credit cannot qualify as biodiesel, renewable diesel, or alternative fuel for purposes of the income tax credit, excise tax credit, or payment provisions relating to those fuels.[7]

Alcohol produced from coal and natural gas

As noted above, the alcohol incentives discussed above do not apply to alcohol produced from petroleum, natural gas, or coal. There is a reduced rate of tax for fuel alcohol produced from natural gas. The reduced rate of tax for fuel alcohol produced from coal has expired.

Reduced rate of tax for alcohol fuels produced from natural gas

A reduced rate of fuel tax is provided for "partially exempt methanol or ethanol fuel." The term "partially exempt methanol or ethanol fuel" means any liquid at least 85 percent of which consists of methanol, ethanol or other alcohol (including methanol and ethanol) produced from natural gas. Partially exempt methanol (or other alcohol) is taxed at 9.15 cents per gallon. Partially exempt ethanol is taxed at 11.3 cents per gallon. After September 30, 2011, these rates drop to 2.15 cents per gallon and 4.3 cents per gallon, respectively.

Reduced rate of tax for alcohol fuels produced from coal

A reduced rate of fuel tax was also provided for "qualified methanol and ethanol fuel." The term "qualified methanol or ethanol fuel" means any liquid at least 85 percent of which consists of methanol, ethanol, or other alcohol produced from coal (including peat). Qualified ethanol was taxed at 13.25 cents per gallon. Qualified methanol was taxed at 12.35 cents per gallon. The incentive expired January 1, 2009.

Biodiesel

The Code provides an income tax credit for biodiesel fuels (the "biodiesel fuels credit").[8] The biodiesel fuels credit is the sum of three credits: (1) the biodiesel mixture credit, (2) the biodiesel credit, and (3) the small agri-biodiesel producer credit. The biodiesel fuels credit is treated as a general business credit. The amount of the biodiesel fuels credit is includable in gross income. The biodiesel fuels credit is coordinated to take into account benefits from the biodiesel excise tax credit and payment provisions discussed below. The credit does not apply to fuel sold or used after December 31, 2009.

Biodiesel is monoalkyl esters of long chain fatty acids derived from plant or animal matter that meet (1) the registration requirements established by the EPA under section 211 of the Clean Air Act (42 U.S.C. sec. 7545) and (2) the requirements of the American Society of

Testing and Materials ("ASTM") D675 1. Agri-biodiesel is biodiesel derived solely from virgin oils including oils from corn, soybeans, sunflower seeds, cottonseeds, canola, crambe, rapeseeds, safflowers, flaxseeds, rice bran, mustard seeds, camelina, or animal fats.

Biodiesel may be taken into account for purposes of the credit only if the taxpayer obtains a certification (in such form and manner as prescribed by the Secretary) from the producer or importer of the biodiesel that identifies the product produced and the percentage of biodiesel and agri-biodiesel in the product.

Biodiesel mixture credit

The biodiesel mixture credit is $1.00 for each gallon of biodiesel (including agri-biodiesel) used by the taxpayer in the production of a qualified biodiesel mixture. A qualified biodiesel mixture is a mixture of biodiesel and diesel fuel that is (1) sold by the taxpayer producing such mixture to any person for use as a fuel, or (2) used as a fuel by the taxpayer producing such mixture. The sale or use must be in the trade or business of the taxpayer and is to be taken into account for the taxable year in which such sale or use occurs. No credit is allowed with respect to any casual off-farm production of a qualified biodiesel mixture.

Per IRS guidance a mixture need only contain 1/10th of one percent of diesel fuel to be a qualified mixture. Thus, a qualified biodiesel mixture can contain 99.9 percent biodiesel and 0.1 percent diesel fuel.

Biodiesel credit (straight biodiesel)

The biodiesel credit is $1.00 for each gallon of biodiesel that is not in a mixture with diesel fuel (100 percent biodiesel or B-100) and which during the taxable year is (1) used by the taxpayer as a fuel in a trade or business or (2) sold by the taxpayer at retail to a person and placed in the fuel tank of such person's vehicle.

Small agri-biodiesel producer credit

The Code provides a small agri-biodiesel producer income tax credit, in addition to the biodiesel and biodiesel fuel mixture credits. The credit is a 10-cents-per-gallon credit for up to 15 million gallons of agri-biodiesel produced by small producers, defined generally as persons whose agri-biodiesel production capacity does not exceed 60 million gallons per year. The agribiodiesel must (1) be sold by such producer to another person (a) for use by such other person in the production of a qualified biodiesel mixture in such person's trade or business (other than casual off-farm production), (b) for use by such other person as a fuel in a trade or business, or, (c) who sells such agri-biodiesel at retail to another person and places such agri-biodiesel in the fuel tank of such other person; or (2) used by the producer for any purpose described in (a), (b), or (c).

Biodiesel mixture excise tax credit

The Code also provides an excise tax credit for biodiesel mixtures.[9] The credit is $1.00 for each gallon of biodiesel used by the taxpayer in producing a biodiesel mixture for sale or use in a trade or business of the taxpayer. A biodiesel mixture is a mixture of biodiesel and diesel fuel that (1) is sold by the taxpayer producing such mixture to any person for use as a fuel or (2) is used as a fuel by the taxpayer producing such mixture. No credit is allowed unless the taxpayer obtains a certification (in such form and manner as prescribed by the

Secretary) from the producer of the biodiesel that identifies the product produced and the percentage of biodiesel and agri-biodiesel in the product.[10]

The credit is not available for any sale or use for any period after December 31, 2009. This excise tax credit is coordinated with the income tax credit for biodiesel such that credit for the same biodiesel cannot be claimed for both income and excise tax purposes.

Payments with respect to biodiesel fuel mixtures

If any person produces a biodiesel fuel mixture in such person's trade or business, the Secretary is to pay such person an amount equal to the biodiesel mixture credit.[11] The biodiesel fuel mixture credit must first be taken against tax liability for taxable fuels. To the extent the biodiesel fuel mixture credit exceeds such tax liability, the excess may be received as a payment. Thus, if the person has no section 4081 liability, the credit is refundable. The Secretary is not required to make payments with respect to biodiesel fuel mixtures sold or used after December 31, 2009.

Renewable diesel

"Renewable diesel" is liquid fuel that (1) is derived from biomass (as defined in section 45K(c)(3)), (2) meets the registration requirements for fuels and fuel additives established by the EPA under section 211 of the Clean Air Act, and (3) meets the requirements of the ASTM D975 or D396, or equivalent standard established by the Secretary. ASTM D975 provides standards for diesel fuel suitable for use in diesel engines. ASTM D396 provides standards for fuel oil intended for use in fuel-oil burning equipment, such as furnaces. Renewable diesel also includes fuel derived from biomass that meets the requirements of a Department of Defense specification for military jet fuel or an ASTM for aviation turbine fuel.

For purposes of the Code, renewable diesel is generally treated the same as biodiesel. In the case of renewable diesel that is aviation fuel, kerosene is treated as though it were diesel fuel for purposes of a qualified renewable diesel mixture. Like biodiesel, the incentive may be taken as an income tax credit, an excise tax credit, or as a payment from the Secretary.[12] The incentive for renewable diesel is $1.00 per gallon. There is no small producer credit for renewable diesel. The incentives for renewable diesel expire after December 31, 2009.

Alternative fuel excise tax credits and payments

The Code provides two per-gallon excise tax credits with respect to alternative fuel, the alternative fuel credit, and the alternative fuel mixture credit. For this purpose, the term "alternative fuel" means liquefied petroleum gas, P Series fuels (as defined by the Secretary of Energy under 42 U.S.C. sec. 13211(2)), compressed or liquefied natural gas, liquefied hydrogen, liquid fuel derived from coal through the Fischer-Tropsch process ("coal-to-liquids"), compressed or liquified gas derived from biomass, or liquid fuel derived from biomass. Such term does not include ethanol, methanol, or biodiesel.

For coal-to-liquids produced after September 30, 2009 through December 30, 2009, the fuel must be certified as having been derived from coal produced at a gasification facility that separates and sequesters 50 percent of such facility's total carbon dioxide emissions. The sequestration percentage increases to 75 percent for fuel produced after December 30, 2009.

The alternative fuel credit is allowed against section 4041 liability, and the alternative fuel mixture credit is allowed against section 4081 liability. Neither credit is allowed unless the taxpayer is registered with the Secretary. The alternative fuel credit is 50 cents per gallon of alternative fuel or gasoline gallon equivalents[13] of nonliquid alternative fuel sold by the taxpayer for use as a motor fuel in a motor vehicle or motorboat, sold for use in aviation or so used by the taxpayer.

The alternative fuel mixture credit is 50 cents per gallon of alternative fuel used in producing an alternative fuel mixture for sale or use in a trade or business of the taxpayer. An "alternative fuel mixture" is a mixture of alternative fuel and taxable fuel that contains at least 1/10 of one percent taxable fuel.[14] The mixture must be sold by the taxpayer producing such mixture to any person for use as a fuel, or used by the taxpayer producing the mixture as a fuel. The credits generally expire after December 31, 2009.

A person may file a claim for payment equal to the amount of the alternative fuel credit and alternative fuel mixture credits. These payment provisions generally also expire after December 31, 2009. With respect to liquefied hydrogen, the credit and payment provisions expire after September 30, 2014. The alternative fuel credit and alternative fuel mixture credit must first be applied to excise tax liability for special and alternative fuels, and any excess credit may be taken as a payment.

3. Renewable electricity production credit (sec. 45)

In general

An income tax credit is allowed for the production of electricity from qualified energy resources at qualified facilities (the "renewable electricity production credit").[15] Qualified energy resources comprise wind, closed-loop biomass, open-loop biomass, geothermal energy, solar energy, small irrigation power, municipal solid waste, qualified hydropower production, and marine and hydrokinetic renewable energy. Qualified facilities are, generally, facilities that generate electricity using qualified energy resources. To be eligible for the credit, electricity produced from qualified energy resources at qualified facilities must be sold by the taxpayer to an unrelated person.

Credit amounts and credit period

In general

The base amount of the electricity production credit is 1.5 cents (indexed annually for inflation) per kilowatt-hour of electricity produced. The amount of the credit was 2.1 cents per kilowatt-hour for 2008. A taxpayer may generally claim a credit during the 10-year period commencing with the date the qualified facility is placed in service. The credit is reduced for grants, tax-exempt bonds, subsidized energy financing, and other credits.

Credit phaseout

The amount of credit a taxpayer may claim is phased out as the market price of electricity exceeds certain threshold levels. The electricity production credit is reduced over a 3-cent phaseout range to the extent the annual average contract price per kilowatt-hour of electricity sold in the prior year from the same qualified energy resource exceeds 8 cents (adjusted for inflation; 11.8 cents for 2008).

Reduced credit periods and credit amounts

Generally, in the case of open-loop biomass facilities (including agricultural livestock waste nutrient facilities), geothermal energy facilities, solar energy facilities, small irrigation power facilities, landfill gas facilities, and trash combustion facilities placed in service before August 8, 2005, the 10-year credit period is reduced to five years, commencing on the date the facility was originally placed in service. However, for qualified open-loop biomass facilities (other than a facility described in section 45(d)(3)(A)(i) that uses agricultural livestock waste nutrients) placed in service before October 22, 2004, the five-year period commences on January 1, 2005. In the case of a closed-loop biomass facility modified to co-fire with coal, to co-fire with other biomass, or to co-fire with coal and other biomass, the credit period begins no earlier than October 22, 2004.

In the case of open-loop biomass facilities (including agricultural livestock waste nutrient facilities), small irrigation power facilities, landfill gas facilities, trash combustion facilities, qualified hydropower facilities, and marine and hydrokinetic renewable energy facilities, the otherwise allowable credit amount is 0.75 cent per kilowatt-hour, indexed for inflation measured after 1992 (1 cent per kilowatt-hour for 2008).

Other limitations on credit claimants and credit amounts

In general, to claim the credit, a taxpayer must own the qualified facility and sell the electricity produced by the facility to an unrelated party. A lessee or operator may claim the credit in lieu of the owner of the qualifying facility in the case of qualifying open-loop biomass facilities and in the case of closed-loop biomass facilities modified to co-fire with coal, to co-fire with other biomass, or to co-fire with coal and other biomass. In the case of a poultry waste facility, the taxpayer may claim the credit as a lessee or operator of a facility owned by a governmental unit.

For all qualifying facilities, other than closed-loop biomass facilities modified to co-fire with coal, to co-fire with other biomass, or to co-fire with coal and other biomass, the amount of credit a taxpayer may claim is reduced by reason of grants, tax-exempt bonds, subsidized energy financing, and other credits, but the reduction cannot exceed 50 percent of the otherwise allowable credit. In the case of closed-loop biomass facilities modified to co-fire with coal, to co-fire with other biomass, or to co-fire with coal and other biomass, there is no reduction in credit by reason of grants, tax-exempt bonds, subsidized energy financing, and other credits.

The credit for electricity produced from renewable resources is a component of the general business credit.[16] Generally, the general business credit for any taxable year may not exceed the amount by which the taxpayer's net income tax exceeds the greater of the tentative minimum tax or 25 percent of so much of the net regular tax liability as exceeds $25,000. However, this limitation does not apply to section 45 credits for electricity or refined coal produced from a facility (placed in service after October 22, 2004) during the first four years of production beginning on the date the facility is placed in service.[17] Excess credits may be carried back one year and forward up to 20 years.

Qualified facilities

Wind energy facility

A wind energy facility is a facility that uses wind to produce electricity. To be a qualified facility, a wind energy facility must be placed in service after December 31, 1993, and before January 1, 2013.

Closed-loop biomass facility

A closed-loop biomass facility is a facility that uses any organic material from a plant that is planted exclusively for the purpose of being used at a qualifying facility to produce electricity. In addition, a facility can be a closed-loop biomass facility if it is a facility that is modified to use closed-loop biomass to co-fire with coal, with other biomass, or with both coal and other biomass, but only if the modification is approved under the Biomass Power for Rural Development Programs or is part of a pilot project of the Commodity Credit Corporation.

To be a qualified facility, a closed-loop biomass facility must be placed in service after December 31, 1992, and before January 1, 2014. In the case of a facility using closed-loop biomass but also co-firing the closed-loop biomass with coal, other biomass, or coal and other biomass, a qualified facility must be originally placed in service and modified to co-fire the closed-loop biomass at any time before January 1, 2014.

A qualified facility includes a new power generation unit placed in service after October 3, 2008, at an existing closed-loop biomass facility, but only to the extent of the increased amount of electricity produced at the existing facility by reason of such new unit.

Open-loop biomass (including agricultural livestock waste nutrients) facility

An open-loop biomass facility is a facility that uses open-loop biomass to produce electricity. For purposes of the credit, open-loop biomass is defined as (1) any agricultural livestock waste nutrients or (2) any solid, nonhazardous, cellulosic waste material or any lignin material that is segregated from other waste materials and which is derived from:

- forest-related resources, including mill and harvesting residues, precommercial thinnings, slash, and brush;
- solid wood waste materials, including waste pallets, crates, dunnage, manufacturing and construction wood wastes, and landscape or right-of-way tree trimmings; or
- agricultural sources, including orchard tree crops, vineyard, grain, legumes, sugar, and other crop by-products or residues.

Agricultural livestock waste nutrients are defined as agricultural livestock manure and litter, including bedding material for the disposition of manure. Wood waste materials do not qualify as open-loop biomass to the extent they are pressure treated, chemically treated, or painted. In addition, municipal solid waste, gas derived from the biodegradation of solid waste, and paper that is commonly recycled do not qualify as open-loop biomass. Open-loop biomass does not include closed-loop biomass or any biomass burned in conjunction with fossil fuel (co- firing) beyond such fossil fuel required for start up and flame stabilization.

In the case of an open-loop biomass facility that uses agricultural livestock waste nutrients, a qualified facility is one that was originally placed in service after October 22, 2004, and before January 1, 2014, and has a nameplate capacity rating which is not less than 150 kilowatts. In the case of any other open-loop biomass facility, a qualified facility is one that was originally placed in service before January 1, 2014. A qualified facility includes a new power generation unit placed in service after October 3, 2008, at an existing open-loop biomass facility, but only to the extent of the increased amount of electricity produced at the existing facility by reason of such new unit.

Geothermal facility

A geothermal facility is a facility that uses geothermal energy to produce electricity. Geothermal energy is energy derived from a geothermal deposit that is a geothermal reservoir consisting of natural heat that is stored in rocks or in an aqueous liquid or vapor (whether or not under pressure). To be a qualified facility, a geothermal facility must be placed in service after October 22, 2004, and before January 1, 2014.

Solar facility

A solar facility is a facility that uses solar energy to produce electricity. To be a qualified facility, a solar facility must be placed in service after October 22, 2004, and before January 1, 2006.

Small irrigation facility

A small irrigation power facility is a facility that generates electric power through an irrigation system canal or ditch without any dam or impoundment of water. The installed capacity of a qualified facility must be at least 150 kilowatts but less than five megawatts. To be a qualified facility, a small irrigation facility must be originally placed in service after October 22, 2004, and before October 3, 2008. Marine and hydrokinetic renewable energy facilities, described below, subsume small irrigation power facilities after October 2, 2008.

Landfill gas facility

A landfill gas facility is a facility that uses landfill gas to produce electricity. Landfill gas is defined as methane gas derived from the biodegradation of municipal solid waste. To be a qualified facility, a landfill gas facility must be placed in service after October 22, 2004, and before January 1, 2014.

Trash combustion facility

Trash combustion facilities are facilities that use municipal solid waste (garbage) to produce steam to drive a turbine for the production of electricity. To be a qualified facility, a trash combustion facility must be placed in service after October 22, 2004, and before January 1, 2014. A qualified trash combustion facility includes a new unit, placed in service after October 22, 2004, that increases electricity production capacity at an existing trash combustion facility. A new unit generally would include a new burner/boiler and turbine. The new unit may share certain common equipment, such as trash handling equipment, with other pre-existing units at the same facility. Electricity produced at a new unit of an existing facility qualifies for the production credit only to the extent of the increased amount of electricity produced at the entire facility.

Hydropower facility

A qualifying hydropower facility is (1) a facility that produced hydroelectric power (a hydroelectric dam) prior to August 8, 2005, at which efficiency improvements or additions to capacity have been made after such date and before January 1, 2011, that enable the taxpayer to produce incremental hydropower or (2) a facility placed in service before August 8, 2005, that did not produce hydroelectric power (a nonhydroelectric dam) on such date, and to which turbines or other electricity generating equipment have been added after such date and before January 1, 2014.

At an existing hydroelectric facility, the taxpayer may claim credit only for the production of incremental hydroelectric power. Incremental hydroelectric power for any taxable year is equal to the percentage of average annual hydroelectric power produced at the facility attributable to the efficiency improvement or additions of capacity determined by using the same water flow information used to determine an historic average annual hydroelectric power production baseline for that facility. The Federal Energy Regulatory Commission will certify the baseline power production of the facility and the percentage increase due to the efficiency and capacity improvements.

Nonhydroelectric dams converted to produce electricity must be licensed by the Federal Energy Regulatory Commission and meet all other applicable environmental, licensing, and regulatory requirements.

For a nonhydroelectric dam converted to produce electric power before January 1, 2009, there must not be any enlargement of the diversion structure, construction or enlargement of a bypass channel, or the impoundment or any withholding of additional water from the natural stream channel.

For a nonhydroelectric dam converted to produce electric power after December 31, 2008, the nonhydroelectric dam must (1) have been placed in service before October 3, 2008, (2) have been operated for flood control, navigation, or water supply purposes and (3) not have produce hydroelectric power on October 3, 2008. In addition, the hydroelectric project must be operated so that the water surface elevation at any given location and time that would have occurred in the absence of the hydroelectric project is maintained, subject to any license requirements imposed under applicable law that change the water surface elevation for the purpose of improving environmental quality of the affected waterway. The Secretary, in consultation with the Federal Energy Regulatory Commission, shall certify if a hydroelectric project licensed at a nonhydroelectric dam meets this criteria.

Marine and hydrokinetic renewable energy facility

A qualified marine and hydrokinetic renewable energy facility is any facility that produces electric power from marine and hydrokinetic renewable energy, has a nameplate capacity rating of at least 150 kilowatts, and is placed in service after October 2, 2008, and before January 1, 2014. Marine and hydrokinetic renewable energy is defined as energy derived from (1) waves, tides, and currents in oceans, estuaries, and tidal areas; (2) free flowing water in rivers, lakes, and streams; (3) free flowing water in an irrigation system, canal, or other manmade channel, including projects that utilize nonmechanical structures to accelerate the flow of water for electric power production purposes; or (4) differentials in ocean temperature (ocean thermal energy conversion). The term does not include energy derived from any source that uses a dam, diversionary structure (except for irrigation systems, canals, and other man-made channels), or impoundment for electric power production.

Taxation of cooperatives and their patrons

For Federal income tax purposes, a cooperative generally computes its income as if it were a taxable corporation, with one exception: the cooperative may exclude from its taxable income distributions of patronage dividends. Generally, a cooperative that is subject to the cooperative tax rules of subchapter T of the Code[18] is permitted a deduction for patronage dividends paid only to the extent of net income that is derived from transactions with patrons who are members of the cooperative.[19] The availability of such deductions from taxable income has the effect of allowing the cooperative to be treated like a conduit with respect to profits derived from transactions with patrons who are members of the cooperative.

Eligible cooperatives may elect to pass any portion of the credit through to their patrons. An eligible cooperative is defined as a cooperative organization that is owned more than 50 percent by agricultural producers or entities owned by agricultural producers. The credit may be apportioned among patrons eligible to share in patronage dividends on the basis of the quantity or value of business done with or for such patrons for the taxable year. The election must be made on a timely filed return for the taxable year and, once made, is irrevocable for such taxable year.

4. Energy credit (sec. 48)

In general

A nonrefundable, 10-percent business energy credit[20] is allowed for the cost of new property that is equipment that either (1) uses solar energy to generate electricity, to heat or cool a structure, or to provide solar process heat or (2) is used to produce, distribute, or use energy derived from a geothermal deposit, but only, in the case of electricity generated by geothermal power, up to the electric transmission stage. Property used to generate energy for the purposes of heating a swimming pool is not eligible solar energy property.

The energy credit is a component of the general business credit.[21] An unused general business credit generally may be carried back one year and carried forward 20 years.[22] The taxpayer's basis in the property is reduced by one-half of the amount of the credit claimed. For projects whose construction time is expected to equal or exceed two years, the credit may be claimed as progress expenditures are made on the project, rather than during the year the property is placed in service. The credit is allowed against the alternative minimum tax for credits determined in taxable years beginning after October 3, 2008.

Special rules for solar energy property

The credit for solar energy property is increased to 30 percent in the case of periods prior to January 1, 2017. Additionally, equipment that uses fiber-optic distributed sunlight to illuminate the inside of a structure is solar energy property eligible for the 30-percent credit.

Fuel cells and microturbines

The energy credit applies to qualified fuel cell power plants, but only for periods prior to January 1, 2017. The credit rate is 30 percent.

A qualified fuel cell power plant is an integrated system composed of a fuel cell stack assembly and associated balance of plant components that (1) converts a fuel into electricity using electrochemical means, and (2) has an electricity-only generation efficiency of greater than 30 percent and a capacity of at least one-half kilowatt. The credit may not exceed $1,500 for each 0.5 kilowatt of capacity.

The energy credit applies to qualifying stationary microturbine power plants for periods prior to January 1, 2017. The credit is limited to the lesser of 10 percent of the basis of the property or $200 for each kilowatt of capacity.

A qualified stationary microturbine power plant is an integrated system comprised of a gas turbine engine, a combustor, a recuperator or regenerator, a generator or alternator, and associated balance of plant components that converts a fuel into electricity and thermal energy. Such system also includes all secondary components located between the existing infrastructure for fuel delivery and the existing infrastructure for power distribution, including equipment and controls for meeting relevant power standards, such as voltage, frequency and power factors. Such system must have an electricity-only generation efficiency of not less than 26 percent at International Standard Organization conditions and a capacity of less than 2,000 kilowatts.

Geothermal heat pump property

The energy credit applies to qualified geothermal heat pump property placed in service prior to January 1, 2017. The credit rate is 10 percent. Qualified geothermal heat pump property is equipment that uses the ground or ground water as a thermal energy source to heat a structure or as a thermal energy sink to cool a structure.

Small wind property

The energy credit applies to qualified small wind energy property placed in service prior to January 1, 2017. The credit rate is 30 percent. Qualified small wind energy property is property that uses a qualified wind turbine to generate electricity. A qualifying wind turbine means a wind turbine of 100 kilowatts of rated capacity or less.

Combined heat and power property

The energy credit applies to combined heat and power ("CHP") property placed in service prior to January 1, 2017. The credit rate is 10 percent.

CHP property is property: (1) that uses the same energy source for the simultaneous or sequential generation of electrical power, mechanical shaft power, or both, in combination with the generation of steam or other forms of useful thermal energy (including heating and cooling applications); (2) that has an electrical capacity of not more than 50 megawatts or a mechanical energy capacity of not more than 67,000 horsepower or an equivalent combination of electrical and mechanical energy capacities; (3) that produces at least 20 percent of its total useful energy in the form of thermal energy that is not used to produce electrical or mechanical power, and produces at least 20 percent of its total useful energy in the form of electrical or mechanical power (or a combination thereof); and (4) the energy efficiency percentage of which exceeds 60 percent. CHP property does not include property used to transport the energy source to the generating facility or to distribute energy produced by the facility.

The otherwise allowable credit with respect to CHP property is reduced to the extent the property has an electrical capacity or mechanical capacity in excess of any applicable limits. Property in excess of the applicable limit (15 megawatts or a mechanical energy capacity of more than 20,000 horsepower or an equivalent combination of electrical and mechanical energy capacities) is permitted to claim a fraction of the otherwise allowable credit. The fraction is equal to the applicable limit divided by the capacity of the property. For example, a 45 megawatt property would be eligible to claim 1 5/45ths, or one third, of the otherwise allowable credit. Again, no credit is allowed if the property exceeds the 50 megawatt or 67,000 horsepower limitations described above.

Additionally, systems whose fuel source is at least 90 percent open-loop biomass and that would qualify for the credit but for the failure to meet the efficiency standard are eligible for a credit that is reduced in proportion to the degree to which the system fails to meet the efficiency standard. For example, a system that would otherwise be required to meet the 60-percent efficiency standard, but which only achieves 30-percent efficiency, would be permitted a credit equal to one-half of the otherwise allowable credit (i.e., a 5-percent credit).

Election of energy credit in lieu of sec. 45 production tax credit

A taxpayer may make an irrevocable election to have certain qualified facilities placed in service in 2009 through 2013 (2012 for wind facilities) be treated as energy property eligible for a 30 percent investment credit under section 48. For this purpose, qualified facilities are facilities otherwise eligible for the section 45[23] production tax credit (other than refined coal, Indian coal, and solar facilities) with respect to which no credit under section 45 has been allowed. A taxpayer electing to treat a facility as energy property may not claim the production credit under section 45.

The eligible basis for the investment credit for taxpayers making this election is the basis of the depreciable (or amortizable) property that would comprise a section 45 credit-eligible facility. For example, in the case of a wind facility, only property eligible for five- year depreciation under section 1 68(e)(3)(b)(vi) is treated as credit-eligible energy property under the election.

5. Clean renewable energy bonds and new clean renewable energy bonds (secs. 54 and 54C)

Both Clean Renewable Energy Bonds ("CREBs") and New Clean Renewable Energy Bonds ("New CREBs") are types of tax credit bonds that can be used to finance certain facilities that produce electricity from certain renewable resources. CREBs are subject to a national volume cap of $1.2 billion[24] and must be issued by December 31, 2009.[25] A taxpayer holding a CREB on a credit allowance date is entitled to a credit against its Federal income taxes. As discussed in more detail below, the annual credit with respect to a CREB is equal to the amount that the Treasury Secretary determines would allow the CREB to be issued at par and without interest.[26]

New CREBs are subject to a national volume cap of $2.4 billion.[27] A taxpayer holding a New CREB on a credit allowance date is entitled to a credit against its Federal income taxes.

As discussed in more detail below, the annual credit with respect to a New CREB is equal to 70 percent of the credit that the Treasury Secretary determines would allow the New CREB to be issued at par and without interest.[28] Because the credit is only 70 percent of the credit that would permit the New CREB to be issued at par, it is assumed that the New CREB will be interest bearing and/or issued at a discount. The tax credit to a holder of a New CREB is treated as interest that is includible in the holder's gross income, and any interest paid on a New CREB is taxable.[29]

CREBs

CREBs are defined as any bond issued by a "qualified issuer" if, in addition to the requirements discussed below, 95 percent or more of the proceeds from the issuance of such bonds are used to finance capital expenditures incurred by "qualified borrowers" for facilities that qualify for the tax credit under section 45 (other than Indian coal production facilities), without regard to the placed-in-service date requirements of that section.[30] The term "qualified issuer" includes (1) governmental bodies (including Indian tribal governments); (2) mutual or cooperative electric companies (described in section 501 (c)(12) or section 1381 (a)(2)(C), or a not-for-profit electric utility which has received a loan or guarantee under the Rural Electrification Act); and (3) clean energy bond lenders.[31] The term "qualified borrower" includes a governmental body (including an Indian tribal government) and a mutual or cooperative electric company.[32]

Projects that may be financed with CREBs include any facility owned by a qualified borrower that is functionally related and subordinate (as determined under Treas. Reg. sec. 1. 103-8(a)(3)) to any qualified facility described in sections 45(d)(1) through (d)(9) (determined without regard to any placed in service date) and owned by such qualified borrower.[33]

The credit rate on a CREB is determined by the Secretary and is to be a rate that permits issuance of CREBs without discount and interest cost to the qualified issuer.[34] The credit accrues quarterly and is includible in gross income (as if it were an interest payment on the bond), and can be claimed against regular income tax liability and alternative minimum tax liability.

CREBs are subject to a maximum maturity limitation.[35] The maximum maturity is the term which the Secretary estimates will result in the present value of the obligation to repay the principal on a CREBs being equal to 50 percent of the face amount of such bond. In addition, the Code requires level amortization of CREBs during the period such bonds are outstanding.

CREBs also are subject to the arbitrage requirements of section 148 that apply to traditional tax-exempt bonds. Principles under section 148 and the regulations thereunder apply for purposes of determining the yield restriction and arbitrage rebate requirements applicable to CREBs.[36]

To qualify as CREBs, the qualified issuer must reasonably expect to and actually spend 95 percent or more of the proceeds of from the issuance of such bonds on qualified projects within the five-year period that begins on the date of issuance.[37] To the extent less than 95 percent of the proceeds are used to finance qualified projects during the five-year spending period, bonds will continue to qualify as CREBs if unspent proceeds are used within 90 days from the end of such five-year period to redeem any "nonqualified bonds." The five-year spending period may be extended by the Secretary upon the qualified issuer's request demonstrating that the failure to satisfy the five-year requirement is due to reasonable cause and the projects will continue to proceed with due diligence.

Allocations of CREB authority were made in 2006 and 2008. The table below reflects those allocations.

New CREBs

Section 107 of the Energy Improvement and Extension Act of 2008 (Division B of Pub. L. No. 110-343) added a new section 54C of the Code for new clean renewable energy bonds ("New CREBs"). New CREBs may be issued by "qualified issuers" to finance "qualified renewable energy facilities." Qualified renewable energy facilities are facilities that: (1) qualify for the tax credit under section 45 (other than Indian coal and refined coal production facilities), without regard to the placed-in-service date requirements of that section; and (2) are owned by a public power provider, governmental body, or cooperative electric company.

The term "qualified issuer" includes: (1) public power providers; (2) a governmental body; (3) cooperative electric companies; (4) a not-for-profit electric utility that has received a loan or guarantee under the Rural Electrification Act; and (5) clean renewable energy bond lenders. The term "public power provider" means a State utility with a service obligation, as such terms are defined in section 217 of the Federal Power Act (as in effect on the date of the enactment of this paragraph). A "governmental body" means any State or Indian tribal government, or any political subdivision thereof. The term "cooperative electric company" means a mutual or cooperative electric company (described in section 501 (c)(12) or section 1381(a)(2)(C)). A clean renewable energy bond lender means a cooperative that is owned by, or has outstanding loans to, 100 or more cooperative electric companies and is in existence on February 1, 2002 (including any affiliated entity which is controlled by such lender).

There is a national limitation for New CREBs of $2.4 billion.[39] No more than one third of the national limit may be allocated to projects of public power providers, governmental bodies, or cooperative electric companies. Allocations to governmental bodies and cooperative electric companies may be made in the manner the Secretary determines appropriate. Allocations to projects of public power providers shall be made, to the extent practicable, in such manner that the amount allocated to each such project bears the same ratio to the cost of such project as the maximum allocation limitation to projects of public power providers bears to the cost of all such projects.

Table 1. Clean Renewable Energy Bond Allocations[38] (2006 and 2008)

Type of Borrower	Type of Project	Number of Projects
Governmental	Solar	539
	Wind	187
	Landfill gas	64
	Open loop biomass	1
	Closed loop biomass	3
	Trash combustion	3
	Hydropower	13
Cooperatives	Solar	34
	Wind	27
	Landfill gas	17
	Open loop biomass	13
	Hydropower	12
	Refined coal	1

New CREBs are a type of qualified tax credit bond for purposes of section 54A of the Code. As such, 100 percent of the available project proceeds of New CREBs must be used within the three-year period that begins on the date of issuance.[40] Available project proceeds are proceeds from the sale of the bond issue less issuance costs (not to exceed two percent) and any investment earnings on such sale proceeds. To the extent less than 100 percent of the available project proceeds are used to finance qualified projects during the three-year spending period, bonds will continue to qualify as New CREBs if unspent proceeds are used within 90 days from the end of such three-year period to redeem bonds. The three-year spending period may be extended by the Secretary upon the qualified issuer's request demonstrating that the failure to satisfy the three-year requirement is due to reasonable cause and the projects will continue to proceed with due diligence.

New CREBs generally are subject to the arbitrage requirements of section 148.[41] However, available project proceeds invested during the three-year spending period are not subject to the arbitrage restrictions (i.e., yield restriction and rebate requirements). In addition, amounts invested in a reserve fund are not subject to the arbitrage restrictions to the extent: (1) such fund is funded at a rate not more rapid than equal annual installments; (2) such fund is funded in a manner reasonably expected to result in an amount not greater than an amount necessary to repay the issue; and (3) the yield on such fund is not greater than the average annual interest rate of tax-exempt obligations having a term of 10 years or more that are issued during the month the New CREBs are issued.

As with other tax credit bonds, a taxpayer holding a New CREB on a credit allowance date is entitled to a tax credit; the credit rate on New CREBs is set by the Secretary at a rate that is 70 percent of the rate that would permit issuance of such bonds without discount and interest cost to the issuer.[42] The Secretary determines credit rates for tax credit bonds based on general assumptions about credit quality of the class of potential eligible issuers and such other factors as the Secretary deems appropriate. The Secretary may determine credit rates based on general credit market yield indexes and credit ratings.[43]

The amount of the tax credit is determined by multiplying the bond's credit rate by the face amount of the holder's bond. The credit accrues quarterly, is includible in gross income (as if it were an interest payment on the bond), and can be claimed against regular income tax liability and alternative minimum tax liability. Unused credits may be carried forward to succeeding taxable years. In addition, credits may be separated from the ownership of the underlying bond similar to how interest coupons can be stripped for interest-bearing bonds.

An issuer of New CREBs is treated as meeting the "prohibition on financial conflicts of interest" requirement in section 54A(d)(6) if it certifies that it satisfies (1) applicable State and local law requirements governing conflicts of interest and (2) any additional conflict of interest rules prescribed by the Secretary with respect to any Federal, State, or local government official directly involved with the issuance of New CREBs.

6. Five-year cost recovery of certain energy property (sec. 168)

In general

A taxpayer is allowed to recover, through annual depreciation deductions, the cost of certain property used in a trade or business or for the production of income. The amount of the depreciation deduction allowed with respect to tangible property for a taxable year is determined under the modified accelerated cost recovery system ("MACRS"). Under MACRS, different types of property are generally assigned applicable recovery periods and depreciation methods. The recovery periods applicable to most tangible personal property (generally tangible property other than residential rental property and nonresidential real property) range from three to 25 years. The depreciation methods generally applicable to tangible personal property are the 200- percent and 150-percent declining balance methods, switching to the straight-line method for the taxable year in which the depreciation deduction would be maximized.[44] In general, the recovery periods applicable to real property are 39 years for non-residential real property and 27.5 years for residential rental property. The depreciation method for real property is the straight-line method.

Under MACRS, the full basis of depreciable property is recovered by the taxpayer over the applicable recovery period; there is no need to estimate salvage value. Furthermore, under MACRS, the applicable recovery period need not (and typically does not) correspond to the actual economic life of the asset subject to depreciation. In general, however, MACRS generally provides for longer recovery periods for longer lived assets.

Certain energy property

The Code provides a five-year statutory recovery period for certain energy property,[45] including (i) equipment which uses solar and wind energy to generate electricity, to heat or cool (or provide hot water for use in) a structure, or to provide solar process heat, except property used to generate energy for the purposes of heating a swimming pool; (ii) equipment which uses solar energy to illuminate the inside of a structure using fiber-optic distributed sunlight;[46] (iii) equipment used to produce, distribute, or use energy derived from a geothermal deposit,[47] but only, in the case of electricity generated by geothermal power, up to (but not including) the electrical transmission stage; and (iv) qualified fuel cell property[48] or qualified microturbine property.[49]

Biomass property

Certain biomass property has a statutory recovery period of five years.[50] For purposes of this provision, biomass property includes (i) a boiler, the primary fuel for which will be an alternate substance; (ii) a burner (including necessary on-site equipment to bring the alternate substance to the burner) for a combustor other than a boiler if the primary fuel for such burner will be an alternate substance; and (iii) equipment for converting an alternate substance into a qualified fuel.[51] Such property also includes pollution control equipment[52] required (by Federal, State, or local regulations) to be installed on or in connection with the above mentioned property and equipment used for the unloading, transfer, storage, reclaiming from storage, and preparation (including, but not limited to, washing, crushing, drying, and weighing) at the point of use of an alternate substance for use in equipment described above.[53]

For purposes of this provision, the term "alternate substance" means any substance other than oil and natural gas, and any product of oil and natural gas.

The property must also be a "qualifying small power production facility." For purposes of this provision, the term "qualifying small power production facility" means a "small power production facility" that the Federal Power Commission determines, by rule, meets such requirements (including requirements respecting fuel use, fuel efficiency, and reliability) as the Commission may, by rule, prescribe, and which is owned by a person not primarily engaged in the generation or sale of electric power (other than electric power solely from cogeneration facilities or small power production facilities).[54] The term "small power production facility" means a facility which (i) produces electric energy solely by the use, as a primary energy source, of biomass, waste, renewable resources, geothermal resources, or any combination thereof; and (ii) has a power production capacity which, together with any other facilities located at the same site (as determined by the Federal Power Commission), is not greater than 80 megawatts.[55]

7. Special allowance for cellulosic biofuel plant property (sec. 168(l))

The Code provides an additional first-year depreciation deduction equal to 50 percent of the adjusted basis of qualified cellulosic biofuel plant property. In order to qualify, the property generally must be placed in service before January 1, 2013.

Qualified cellulosic biofuel plant property means property used in the U.S. solely to produce cellulosic biofuel. For this purpose, cellulosic biofuel means any liquid fuel which is produced by any lignocellulosic or hemicellulosic matter that is available on a renewable or recurring basis. For example, lignocellulosic or hemicellulosic matter that is available on a renewable or recurring basis includes bagasse (from sugar cane), corn stalks, and switchgrass.

The additional first-year depreciation deduction is allowed for both regular tax and alternative minimum tax purposes for the taxable year in which the property is placed in service. The additional first-year depreciation deduction is subject to the general rules regarding whether an item is deductible under section 162 or subject to capitalization under section 263 or section 263A. The basis of the property and the depreciation allowances in the year of purchase and later years are appropriately adjusted to reflect the additional first-year depreciation deduction. In addition, the provision provides that there is no adjustment to the allowable amount of depreciation for purposes of computing a taxpayer's alternative minimum taxable income with respect to property to which the provision applies. A taxpayer is allowed to elect out of the additional first-year depreciation for any class of property for any taxable year.

To qualify for the additional first-year depreciation deduction, the original use of the property must commence with the taxpayer after December 20, 2006. The property must be acquired by purchase (as defined under section 179(d)) by the taxpayer after December 20, 2006 and be placed in service before January 1, 2013. Property does not qualify if a binding written contract for the acquisition of such property was in effect on or before December 20, 2006.

Property that is manufactured, constructed, or produced by the taxpayer for use by the taxpayer qualifies if the taxpayer begins the manufacture, construction, or production of the property after December 20, 2006, and the property is placed in service before January 1, 2013 (and all other requirements are met). Property that is manufactured, constructed, or produced for the taxpayer by another person under a contract that is entered into prior to the manufacture, construction, or production of the property is considered to be manufactured, constructed, or produced by the taxpayer.

Qualified cellulosic biofuel plant property does not include property to which section 168(k) (providing "bonus depreciation") applies; alternative depreciation property as defined under section 168(k)(2)(D)(i); tax-exempt bond financed property; or property for which the taxpayer has elected not to claim the special allowance under section 168(l). Recapture rules apply if the property ceases to be qualified cellulosic biofuel plant property.

B. Provisions Relating to Energy Conservation

1. Credit for nonbusiness energy property (sec. 25C)

In general

Section 25C provides a 30-percent credit for the purchase of qualified energy efficiency improvements to the envelope of existing homes. Additionally, section 25C provides a 30 percent credit for the purchase of (1) qualified natural gas, propane, or oil furnace or hot water boilers, (2) qualified energy efficient property, and (3) advanced main air circulating fans.

The credit applies to expenditures made after December 31, 2008 for property placed in service after December 31, 2008, and prior to January 1, 2011.[56] The aggregate amount of the credit allowed for a taxpayer for taxable years beginning in 2009 and 2010 is $1,500.

Building envelope improvements

A qualified energy efficiency improvement is any energy efficiency building envelope component (1) that meets or exceeds the prescriptive criteria for such a component established by the 2000 International Energy Conservation Code[57] as supplemented and as in effect on August 8, 2005 (or, in the case of metal roofs with appropriate pigmented coatings, meets the Energy Star program requirements); (2) that is installed in or on a dwelling located in the United States and owned and used by the taxpayer as the taxpayer's principal residence; (3) the original use of which commences with the taxpayer; and (4) that reasonably can be expected to remain in use for at least five years. The credit is nonrefundable.

Building envelope components are: (1) insulation materials or systems which are specifically and primarily designed to reduce the heat loss or gain for a dwelling and which meet the prescriptive criteria for such material or system established by the 2009 International Energy Conservation Code, as such Code (including supplements) is in effect on the date of the enactment of the American Recovery and Reinvestment Tax Act of 2009 (February 17, 2009); (2) exterior windows (including skylights) and doors provided such component has a U-factor and a seasonal heat gain coefficient ("SHGC") of 0.3 or less; and (3) metal or asphalt roofs with appropriate pigmented coatings or cooling granules that are specifically and primarily designed to reduce the heat gain for a dwelling.

Other eligible property

Qualified natural gas, propane, or oil furnace or hot water boilers

A qualified natural gas, propane, or oil hot water boiler is a natural gas, propane, or oil hot water boiler with an annual fuel utilization efficiency rate of at least 90. A qualified natural gas or propane furnace is a natural gas or propane furnace with an annual fuel utilization efficiency rate of at least 95. A qualified oil furnace is an oil furnace with an annual fuel utilization efficiency rate of at least 90.

Qualified energy-efficient property

Qualified energy-efficient property is: (1) an electric heat pump water heater which yields an energy factor of at least 2.0 in the standard Department of Energy test procedure, (2) an electric heat pump which achieves the highest efficiency tier of Consortium for Energy Efficiency, as in effect on January 1, 2009,[58] (3) a central air conditioner with energy efficiency of at least the highest efficiency tier established by the Consortium for Energy Efficiency as in effect on Jan. 1, 2009[59], (4) a natural gas, propane, or oil water heater which has an energy factor of at least 0.82 or thermal efficiency of at least 90 percent, and (5) biomass fuel property.

Biomass fuel property is a stove that burns biomass fuel to heat a dwelling unit located in the United States and used as a principal residence by the taxpayer, or to heat water for such dwelling unit, and that has a thermal efficiency rating of at least 75 percent as measured using a lower heating value. Biomass fuel is any plant-derived fuel available on a renewable or recurring basis, including agricultural crops and trees, wood and wood waste and residues (including wood pellets), plants (including aquatic plants, grasses, residues, and fibers).

Advanced main air circulating fan

An advanced main air circulating fan is a fan used in a natural gas, propane, or oil furnace originally placed in service by the taxpayer during the taxable year, and which has an annual electricity use of no more than two percent of the total annual energy use of the furnace (as determined in the standard Department of Energy test procedures).

Additional rules

The taxpayer's basis in the property is reduced by the amount of the credit. Special proration rules apply in the case of jointly owned property, condominiums, and tenant-stockholders in cooperative housing corporations. If less than 80 percent of the property is used for nonbusiness purposes, only that portion of expenditures that is used for nonbusiness purposes is taken into account.

2. Credit for residential energy efficient property (sec. 25D)

In general

Section 25D provides a personal tax credit for the purchase of qualified solar electric property and qualified solar water heating property that is used exclusively for purposes other than heating swimming pools and hot tubs. The credit is equal to 30 percent of qualifying expenditures.

Section 25D also provides a 30 percent credit for the purchase of qualified geothermal heat pump property, qualified small wind energy property, and qualified fuel cell power plants. The credit for any fuel cell may not exceed $500 for each 0.5 kilowatt of capacity.

The credit is nonrefundable. The credit with respect to all qualifying property may be claimed against the alternative minimum tax.

The credit applies to property placed in service prior to January 1, 2017

Qualified property

Qualified solar electric property is property that uses solar energy to generate electricity for use in a dwelling unit. Qualifying solar water heating property is property used to heat water for use in a dwelling unit located in the United States and used as a residence if at least half of the energy used by such property for such purpose is derived from the sun.

A qualified fuel cell power plant is an integrated system comprised of a fuel cell stack assembly and associated balance of plant components that (1) converts a fuel into electricity using electrochemical means, (2) has an electricity-only generation efficiency of greater than 30 percent, and (3) has a nameplate capacity of at least one-half kilowatt. The qualified fuel cell power plant must be installed on or in connection with a dwelling unit located in the United States and used by the taxpayer as a principal residence.

Qualified small wind energy property is property that uses a wind turbine to generate electricity for use in a dwelling unit located in the U.S. and used as a residence by the taxpayer.

Qualified geothermal heat pump property means any equipment which (1) uses the ground or ground water as a thermal energy source to heat the dwelling unit or as a thermal energy sink to cool such dwelling unit, (2) meets the requirements of the Energy Star program which are in effect at the time that the expenditure for such equipment is made, and (3) is installed on or in connection with a dwelling unit located in the United States and used as a residence by the taxpayer.

Additional rules

The depreciable basis of the property is reduced by the amount of the credit. Expenditures for labor costs allocable to onsite preparation, assembly, or original installation of property eligible for the credit are eligible expenditures.

Special proration rules apply in the case of jointly owned property, condominiums, and tenant-stockholders in cooperative housing corporations. If less than 80 percent of the property is used for nonbusiness purposes, only that portion of expenditures that is used for nonbusiness purposes is taken into account.

3. Alternative and plug-in electric-drive motor vehicle credits (secs. 30, 30B, and 30D)

Present Law

Alternative motor vehicle credit (sec. 30B)

A credit is available for each new qualified fuel cell vehicle, hybrid vehicle, advanced lean burn technology vehicle, and alternative fuel vehicle placed in service by the taxpayer during the taxable year.[60] In general, the credit amount varies depending upon the type of technology used, the weight class of the vehicle, the amount by which the vehicle exceeds certain fuel economy standards, and, for some vehicles, the estimated lifetime fuel savings. The credit generally is available for vehicles purchased after 2005. The credit terminates after 2009, 2010, or 2014, depending on the type of vehicle.

Table 2. Base Credit Amount for Fuel Cell Vehicles

Vehicle Gross Weight Rating (pounds)	Credit Amount
Vehicle $\leq$ 8,500	$8,000
8,500 $<$ vehicle $\leq$ 14,000	$10,000
14,000 $<$ vehicle $\leq$ 26,000	$20,000
26,000 $<$ vehicle	$40,000

In general, the credit is allowed to the vehicle owner, including the lessor of a vehicle subject to a lease. If the use of the vehicle is described in paragraphs (3) or (4) of section 50(b) (relating to use by tax-exempt organizations, governments, and foreign persons) and is not subject to a lease, the seller of the vehicle may claim the credit so long as the seller clearly discloses to the user in a document the amount that is allowable as a credit. A vehicle must be used predominantly in the United States to qualify for the credit. The portion of the credit attributable to vehicles of a character subject to an allowance for depreciation is treated as a portion of the general business credit.

Fuel cell vehicles

A qualified fuel cell vehicle is a motor vehicle that is propelled by power derived from one or more cells that convert chemical energy directly into electricity by combining oxygen with hydrogen fuel that is stored on board the vehicle and may or may not require reformation prior to use. A qualified fuel cell vehicle must be purchased before January 1, 2015. The amount of credit for the purchase of a fuel cell vehicle is determined by a base credit amount that depends upon the weight class of the vehicle and, in the case of automobiles or light trucks, an additional credit amount that depends upon the rated fuel economy of the vehicle compared to a base fuel economy. For these purposes the base fuel economy is the 2002 model year city fuel economy rating for vehicles of various weight classes.[61] Table 2, below, shows the base credit amounts.

In the case of a fuel cell vehicle weighing less than 8,500 pounds and placed in service after December 31, 2009, the $8,000 amount in Table 2, above is reduced to $4,000.

Table 3, below, shows the additional credits for passenger automobiles or light trucks.

Hybrid vehicles and advanced lean burn technology vehicles

Qualified hybrid vehicles

A qualified hybrid vehicle is a motor vehicle that draws propulsion energy from on-board sources of stored energy that include both an internal combustion engine or heat engine using combustible fuel and a rechargeable energy storage system (e.g., batteries). A qualified hybrid vehicle must be placed in service before January 1, 2011 (January 1, 2010 in the case of a hybrid vehicle weighing more than 8,500 pounds)

Table 3. Credit for Qualified Fuel Cell Vehicles

Credit	If Fuel Economy of the Fuel Cell Vehicle Is:	
	at least	but less than
$1,000	150% of base fuel economy	175% of base fuel economy
$1,500	175% of base fuel economy	200% of base fuel economy
$2,000	200% of base fuel economy	225% of base fuel economy
$2,500	225% of base fuel economy	250% of base fuel economy
$3,000	250% of base fuel economy	275% of base fuel economy
$3,500	275% of base fuel economy	300% of base fuel economy
$4,000	300% of base fuel economy	

Hybrid vehicles that are automobiles and light trucks

In the case of an automobile or light truck (vehicles weighing 8,500 pounds or less), the amount of credit for the purchase of a hybrid vehicle is the sum of two components: (1) a fuel economy credit amount that varies with the rated fuel economy of the vehicle compared to a 2002 model year standard and (2) a conservation credit based on the estimated lifetime fuel savings of the qualified vehicle compared to a comparable 2002 model year vehicle that is powered solely by a gasoline or diesel internal combustion engine. A qualified hybrid automobile or light truck must have a maximum available power[62] from the rechargeable energy storage system of at least four percent. In addition, the vehicle must meet or exceed certain Environmental Protection Agency ("EPA") emissions standards. For a vehicle with a gross vehicle weight rating of 6,000 pounds or less, the applicable emissions standards are the Bin 5 Tier II emissions standards. For a vehicle with a gross vehicle weight rating greater than 6,000 pounds and less than or equal to 8,500 pounds, the applicable emissions standards are the Bin 8 Tier II emissions standards.

Table 4, below, shows the fuel economy credit available to a hybrid passenger automobile or light truck whose fuel economy (on a gasoline gallon equivalent basis) exceeds that of a base fuel economy.

Table 5, below, shows the conservation credit.

Advanced lean burn technology vehicles

The amount of credit for the purchase of an advanced lean burn technology vehicle is the sum of two components: (1) a fuel economy credit amount that varies with the rated fuel economy of the vehicle compared to a 2002 model year standard as described in Table 4, above, and (2) a conservation credit based on the estimated lifetime fuel savings of a qualified vehicle compared to a comparable 2002 model year vehicle as described in Table 5, above. The amounts of the credits are determined after an adjustment is made to account for the different BTU content of gasoline and the fuel utilized by the lean burn technology vehicle.

Table 4. Fuel Economy Credit

Credit	If Fuel Economy of the Hybrid Vehicle Is:	
	at least	**but less than**
\$400	125% of base fuel economy	150% of base fuel economy
\$800	150% of base fuel economy	175% of base fuel economy
\$1,200	175% of base fuel economy	200% of base fuel economy
\$1,600	200% of base fuel economy	225% of base fuel economy
\$2,000	225% of base fuel economy	250% of base fuel economy
\$2,400	250% of base fuel economy	

Table 5. Conservation Credit

Estimated Lifetime Fuel Savings (gallons of gasoline)	Conservation Amount
At least 1,200 but less than 1,800	\$250
At least 1,800 but less than 2,400	\$500
At least 2,400 but less than 3,000	\$750
At least 3,000	\$1,000

A qualified advanced lean burn technology vehicle is a passenger automobile or a light truck that incorporates direct injection, achieves at least 125 percent of the 2002 model year city fuel economy, and for 2004 and later model vehicles meets or exceeds certain Environmental Protection Agency emissions standards. For a vehicle with a gross vehicle weight rating of 6,000 pounds or less the applicable emissions standards are the Bin 5 Tier II emissions standards. For a vehicle with a gross vehicle weight rating greater than 6,000 pounds and less than or equal to 8,500 pounds, the applicable emissions standards are the Bin 8 Tier II emissions standards. A qualified advanced lean burn technology vehicle must be placed in service before January 1, 2011.

Limitation on number of qualified hybrid and advanced lean burn technology vehicles eligible for the credit

There is a limitation on the number of passenger and light truck qualified hybrid vehicles and advanced lean burn technology vehicles sold by each manufacturer of such vehicles that are eligible for the credit. Taxpayers may claim the full amount of the allowable credit up to the end of the first calendar quarter after the quarter in which the manufacturer records the 60,000th hybrid and advanced lean burn technology vehicle sale occurring after December 31, 2005. Taxpayers may claim one half of the otherwise allowable credit during the two calendar quarters subsequent to the first quarter after the manufacturer has recorded its 60,000th such sale. In the third and fourth calendar quarters subsequent to the first quarter after the manufacturer has recorded its 60,000th such sale, the taxpayer may claim one quarter of the otherwise allowable credit.

Thus, for example, summing the sales of qualified hybrid vehicles that are passenger vehicles or light trucks and all sales of qualified advanced lean burn technology vehicles, if a manufacturer records the sale of its 60,000th qualified vehicle in February of 2007, taxpayers purchasing such vehicles from the manufacturer may claim the full amount of the credit on their purchases of qualified vehicles through June 30, 2007. For the period July 1, 2007, through December 31, 2007, taxpayers may claim one half of the otherwise allowable credit on purchases of qualified vehicles of the manufacturer. For the period January 1, 2008, through June 30, 2008, taxpayers may claim one quarter of the otherwise allowable credit on the purchases of qualified vehicles of the manufacturer. After June 30, 2008, no credit may be claimed for purchases of such hybrid vehicles or advanced lean burn technology vehicles sold by the manufacturer.

Hybrid vehicles that are medium and heavy trucks

In the case of a qualified hybrid vehicle weighing more than 8,500 pounds, the amount of credit is determined by the estimated increase in fuel economy and the incremental cost of the hybrid vehicle compared to a comparable vehicle powered solely by a gasoline or diesel internal combustion engine and that is comparable in weight, size, and use of the vehicle. For a vehicle that achieves a fuel economy increase of at least 30 percent but less than 40 percent, the credit is equal to 20 percent of the incremental cost of the hybrid vehicle. For a vehicle that achieves a fuel economy increase of at least 40 percent but less than 50 percent, the credit is equal to 30 percent of the incremental cost of the hybrid vehicle. For a vehicle that achieves a fuel economy increase of 50 percent or more, the credit is equal to 40 percent of the incremental cost of the hybrid vehicle.

The credit is subject to certain maximum applicable incremental cost amounts. For a qualified hybrid vehicle weighing more than 8,500 pounds but not more than 14,000 pounds, the maximum allowable incremental cost amount is $7,500. For a qualified hybrid vehicle weighing more than 14,000 pounds but not more than 26,000 pounds, the maximum allowable incremental cost amount is $15,000. For a qualified hybrid vehicle weighing more than 26,000 pounds, the maximum allowable incremental cost amount is $30,000.

A qualified hybrid vehicle weighing more than 8,500 pounds but not more than 14,000 pounds must have a maximum available power from the rechargeable energy storage system of at least 10 percent. A qualified hybrid vehicle weighing more than 14,000 pounds must have a maximum available power from the rechargeable energy storage system of at least 15 percent.[63]

Alternative fuel vehicle

The credit for the purchase of a new alternative fuel vehicle is 50 percent of the incremental cost of such vehicle, plus an additional 30 percent if the vehicle meets certain emissions standards. The incremental cost of any new qualified alternative fuel vehicle is the excess of the manufacturer's suggested retail price for such vehicle over the price for a gasoline or diesel fuel vehicle of the same model. To be eligible for the credit, a qualified alternative fuel vehicle must be purchased before January 1, 2011.

The amount of the credit varies depending on the weight of the qualified vehicle. The credit is subject to certain maximum applicable incremental cost amounts. Table 6, below, shows the maximum permitted incremental cost for the purpose of calculating the credit for alternative fuel vehicles by vehicle weight class as well as the maximum credit amount for such vehicles.

Alternative fuels comprise compressed natural gas, liquefied natural gas, liquefied petroleum gas, hydrogen, and any liquid fuel that is at least 85 percent methanol. Qualified alternative fuel vehicles are vehicles that operate only on qualified alternative fuels and are incapable of operating on gasoline or diesel (except to the extent gasoline or diesel fuel is part of a qualified mixed fuel, described below).

Certain mixed fuel vehicles, that is vehicles that use a combination of an alternative fuel and a petroleum-based fuel, are eligible for a reduced credit. If the vehicle operates on a mixed fuel that is at least 75 percent alternative fuel, the vehicle is eligible for 70 percent of the otherwise allowable alternative fuel vehicle credit. If the vehicle operates on a mixed fuel that is at least 90 percent alternative fuel, the vehicle is eligible for 90 percent of the otherwise allowable alternative fuel vehicle credit.

Table 6. Maximum Allowable Incremental Cost for Calculation of Alternative Fuel Vehicle Credit

Vehicle Gross Weight Rating (pounds)	Maximum Allowable Incremental Cost	Maximum Allowable Credit
Vehicle $\leq$ 8,500	$5,000	$4,000
8,500 < vehicle $\leq$ 14,000	$10,000	$8,000
14,000 < vehicle $\leq$ 26,000	$25,000	$20,000
26,000 < vehicle	$40,000	$32,000

Table 7. 2002 Model Year City Fuel Economy

Vehicle Inertia Weight Class (pounds)	Passenger Automobile (miles per gallon)	Light Truck (miles per gallon)
1,500	45.2	39.4
1,750	45.2	39.4
2,000	39.6	35.2
2,250	35.2	31.8
2,500	31.7	29.0
2,750	28.8	26.8
3,000	26.4	24.9
3,500	22.6	21.8
4,000	19.8	19.4
4,500	17.6	17.6
5,000	15.9	16.1
5,500	14.4	14.8
6,000	13.2	13.7
6,500	12.2	12.8
7,000	11.3	12.1
8,500	11.3	12.1

Base fuel economy

The base fuel economy is the 2002 model year city fuel economy by vehicle type and vehicle inertia weight class. For this purpose, "vehicle inertia weight class" has the same meaning as when defined in regulations prescribed by the EPA for purposes of Title II of the Clean Air Act. Table 7, below, shows the 2002 model year city fuel economy for vehicles by type and by inertia weight class.

Plug-in electric-drive motor vehicle credit (sec. 30D)

Rules for taxable years beginning after December 31, 2008, with respect to vehicles acquired in that taxable year and before January 1, 2010

A credit is available for each qualified plug-in electric-drive motor vehicle placed in service. A qualified plug-in electric-drive motor vehicle is a motor vehicle that has at least four wheels, is manufactured for use on public roads, meets certain emissions standards (except for certain heavy vehicles), draws propulsion using a traction battery with at least four kilowatt- hours of capacity, and is capable of being recharged from an external source of electricity.

The base amount of the plug-in electric-drive motor vehicle credit is $2,500, plus another $417 for each kilowatt-hour of battery capacity in excess of four kilowatt-hours. The maximum credit for qualified vehicles weighing 10,000 pounds or less is $7,500. This maximum amount increases to $10,000 for vehicles weighing more than 10,000 pounds but not more than 14,000 pounds, to $12,500 for vehicles weighing more than 14,000 pounds but not more than 26,000 pounds, and to $15,000 for vehicle weighing more than 26,000 pounds.

In general, the credit is available to the vehicle owner, including the lessor of a vehicle subject to lease. If the qualified vehicle is used by certain tax-exempt organizations, governments, or foreign persons and is not subject to a lease, the seller of the vehicle may claim the credit so long as the seller clearly discloses to the user in a document the amount that is allowable as a credit. A vehicle must be used predominantly in the United States to qualify for the credit.

Once a total of 250,000 credit-eligible vehicles have been sold for use in the United States, the credit phases out over four calendar quarters. The phaseout period begins in the second calendar quarter following the quarter during which the vehicle cap has been reached. Taxpayers may claim one-half of the otherwise allowable credit during the first two calendar quarters of the phaseout period and twenty-five percent of the otherwise allowable credit during the next two quarters. After this, no credit is available.

The basis of any qualified vehicle is reduced by the amount of the credit. To the extent a vehicle is eligible for credit as a qualified plug-in electric-drive motor vehicle, it is not eligible for credit as a qualified hybrid vehicle under section 30B. The portion of the credit attributable to vehicles of a character subject to an allowance for depreciation is treated as part of the general business credit; the nonbusiness portion of the credit is allowable to the extent of the excess of the regular tax over the alternative minimum tax (reduced by certain other credits) for the taxable year.

Rules for vehicles acquired after December 31, 2009

For plug-in electric drive vehicles acquired after December 31, 2009, the maximum credit is capped at $7,500 regardless of vehicle weight. In addition, after that date no credit is available for low speed plug-in vehicles or for plug-in vehicles weighing 14,000 pounds or more.

After December 31, 2009, the 250,000 total plug-in vehicle limitation is replaced with a 200,000 plug-in vehicles per manufacturer limitation. Under the new limitation, the credit phases out over four calendar quarters beginning in the second calendar quarter following the quarter in which the manufacturer limit is reached.

Credit for electric-drive low-speed vehicles, motorcycles, and three-wheeled vehicles (sec. 30)

A 10-percent credit for low-speed vehicles, motorcycles, and three-wheeled vehicles is available for vehicles that, generally, would otherwise meet the criteria of a qualified plug-in electric-drive motor vehicle but for the fact that they are low-speed vehicles or do not have at least four wheels. Two or three wheeled vehicles must have a battery capacity of at least 2.5 kilowatt-hours. Other vehicles must have a battery capacity of at least 4 kilowatt-hours. The maximum credit for such vehicles is $2,500. The credit is part of the general business credit. The credit is available for vehicles acquired after February 17, 2009 (the date of enactment of the American Recovery and Reinvestment Act of 2009), and before January 1, 2012.

Credit for converting a vehicle into a plug-in electric-drive motor vehicle (sec. 30B)

A 10-percent credit, up to $4,000, is available for the cost of converting any motor vehicle into a qualified plug-in electric-drive motor vehicle. To be eligible for the credit, a qualified plug-in traction battery module must have a capacity of at least 4 kilowatt-hours.

The credit is available for conversions made after February 17, 2009 (the date of enactment of the American Recovery and Reinvestment Act of 2009), and before January 1, 2012.

4. New energy efficient home credit (sec 45L of the Code)

The new energy efficient home credit is available to an eligible contractor for the construction of a qualified new energy-efficient home. To qualify as a new energy-efficient home, the home must be: (1) a dwelling located in the United States, (2) substantially completed after August 8, 2005, and (3) certified in accordance with guidance prescribed by the Secretary to achieve either a 30-percent or 50-percent reduction in heating and cooling energy consumption compared to a comparable dwelling constructed in accordance with the standards of chapter 4 of the 2003 International Energy Conservation Code as in effect (including supplements) on August 8, 2005, and any applicable Federal minimum efficiency standards for heating and cooling equipment.

The credit equals $1,000 in the case of a new home that meets the 30 percent standard and $2,000 in the case of a new home that meets the 50 percent standard.

With respect to homes that meet the 30-percent standard, one-third of such 30 percent savings must come from the building envelope, and with respect to homes that meet the 50-percent standard, one-fifth of such 50 percent savings must come from the building envelope.

Only manufactured homes are eligible for the $1,000 credit. In lieu of meeting the 30 percent efficiency improvement relative to the standards of chapter 4 of the 2003 International Energy Conservation Code, manufactured homes certified by a method prescribed by the Administrator of the Environmental Protection Agency under the Energy Star Labeled Homes program are eligible for the $1,000 credit provided criteria (1) and (2), above, are met.

Manufactured homes are homes that conform to Federal manufactured home construction and safety standards. The eligible contractor is the person who constructed the home, or in the case of a manufactured home, the producer of such home. The credit is part of the general business credit.

The credit applies to homes acquired prior to January 1, 2010.

5. Energy efficient appliance credit (sec. 45M of the Code)

In general

A credit is allowed for the eligible production of certain energy-efficient dishwashers, clothes washers, and refrigerators. The credit is part of the general business credit.

The credits are as follows:

Dishwashers

$45 in the case of a dishwasher that is manufactured in calendar year 2008 or 2009 that uses no more than 324 kilowatt hours per year and 5.8 gallons per cycle, and

$75 in the case of a dishwasher that is manufactured in calendar year 2008, 2009, or 2010 and that uses no more than 307 kilowatt hours per year and 5.0 gallons per cycle (5.5 gallons per cycle for dishwashers designed for greater than 12 place settings).

Clothes washers

$75 in the case of a residential top-loading clothes washer manufactured in calendar year 2008 that meets or exceeds a 1.72 modified energy factor and does not exceed a 8.0 water consumption factor, and

$125 in the case of a residential top-loading clothes washer manufactured in calendar year 2008 or 2009 that meets or exceeds a 1.8 modified energy factor and does not exceed a 7.5 water consumption factor,

$150 in the case of a residential or commercial clothes washer manufactured in calendar year 2008, 2009 or 2010 that meets or exceeds a 2.0 modified energy factor and does not exceed a 6.0 water consumption factor, and

$250 in the case of a residential or commercial clothes washer manufactured in calendar year 2008, 2009, or 2010 that meets or exceeds a 2.2 modified energy factor and does not exceed a 4.5 water consumption factor.

Refrigerators

$50 in the case of a refrigerator manufactured in calendar year 2008 that consumes at least 20 percent but not more than 22.9 percent less kilowatt hours per year than the 2001 energy conservation standards,

$75 in the case of a refrigerator that is manufactured in calendar year 2008 or 2009 that consumes at least 23 percent but no more than 24.9 percent less kilowatt hours per year than the 2001 energy conservation standards,

$100 in the case of a refrigerator that is manufactured in calendar year 2008, 2009 or 2010 that consumes at least 25 percent but not more than 29.9 percent less kilowatt hours per year than the 2001 energy conservation standards, and

$200 in the case of a refrigerator manufactured in calendar year 2008, 2009 or 2010 that consumes at least 30 percent less energy than the 2001 energy conservation standards.

Definitions

A dishwasher is any residential dishwasher subject to the energy conservation standards established by the Department of Energy. A refrigerator must be an automatic defrost refrigerator-freezer with an internal volume of at least 16.5 cubic feet to qualify for the credit. A clothes washer is any residential clothes washer, including a residential style coin operated washer, that satisfies the relevant efficiency standard.

The term "modified energy factor" means the modified energy factor established by the Department of Energy for compliance with the Federal energy conservation standard.

The term "gallons per cycle" means, with respect to a dishwasher, the amount of water, expressed in gallons, required to complete a normal cycle of a dishwasher.

The term "water consumption factor" means, with respect to a clothes washer, the quotient of the total weighted per-cycle water consumption divided by the cubic foot (or liter) capacity of the clothes washer.

Other rules

Appliances eligible for the credit include only those produced in the United States and that exceed the average amount of U.S. production from the two prior calendar years for each category of appliance. The aggregate credit amount allowed with respect to a taxpayer for all taxable years beginning after December 31, 2007 may not exceed $75 million, with the exception that the $200 refrigerator credit and the $250 clothes washer credit are not limited. Additionally, the credit allowed in a taxable year for all appliances may not exceed two percent of the average annual gross receipts of the taxpayer for the three taxable years preceding the taxable year in which the credit is determined.

6. Qualified energy conservation bonds (sec. 54D)

Qualified Energy Conservation Bonds ("QEC bonds") are a type of qualified tax credit bond for purposes of section 54A of the Code.[64] QEC bonds, which are subject to a national volume cap of $3.2 billion[65] (discussed below) generally must be issued by a State or local government and must be designated by the issuer as a QEC bond.[66] In addition, 100 percent of the available project proceeds of QEC bond issuance must be used for "qualified conservation purposes" (defined below).[67] A taxpayer holding a QEC bond on a credit allowance date is entitled to a credit against its Federal income taxes. As discussed in more detail below, the annual credit with respect to a QEC bond is equal to 70 percent of the credit that the Treasury Secretary determines would allow the QEC bond to be issued at par and without interest. Because the credit is only 70 percent of the credit that would permit the QEC bond to be issued at par and without interest, it is assumed that QEC bonds will be interest bearing and/or issued at a discount. The tax credit to a holder of a QEC bond is treated as interest that is includible in the holder's gross income, and any interest paid on a QEC bond is taxable.[68]

The term "qualified conservation purpose" means:

1. Capital expenditures incurred for purposes of (a) reducing energy consumption in publicly owned buildings by at least 20 percent; (b) implementing green community programs (including the use of loans, grants or other repayment mechanisms to implement such programs); (c) rural development involving the production of electricity from renewable energy resources; or (d) any facility eligible for the production tax credit under section 45 (other than Indian coal and refined coal production facilities);[69]
2. Expenditures with respect to facilities or grants that support research in: (a) development of cellulosic ethanol or other nonfossil fuels; (b) technologies for the capture and sequestration of carbon dioxide produced through the use of fossil fuels; (c) increasing the efficiency of existing technologies for producing nonfossil fuels; (d) automobile battery technologies and other technologies to reduce fossil fuel consumption in transportation; and (e) technologies to reduce energy use in buildings;[70]
3. Mass commuting facilities and related facilities that reduce the consumption of energy, including expenditures to reduce pollution from vehicles used for mass commuting;[71]
4. Demonstration projects designed to promote the commercialization of: (a) green building technology; (b) conversion of agricultural waste for use in the production of fuel or otherwise; (c) advanced battery manufacturing technologies; (d) technologies to reduce peak-use of electricity; and (e) technologies for the capture and sequestration of carbon dioxide emitted from combusting fossil fuels in order to produce electricity;[72] and

5. Public education campaigns to promote energy efficiency (other than movies, concerts, and other events held primarily for entertainment purposes).[73]

There is a national limitation on QEC bonds of $3.2 billion.[74] Allocations of QEC bonds are made to the States with sub-allocations to large local governments.[75] Allocations are made to the States according to their respective populations, reduced by any sub-allocations to large local governments (defined below) within the States. Sub-allocations to large local governments shall be an amount of the national qualified energy conservation bond limitation that bears the same ratio to the amount of such limitation that otherwise would be allocated to the State in which such large local government is located as the population of such large local government bears to the population of such State. The term "large local government" means: any municipality or county if such municipality or county has a population of 100,000 or more. Indian tribal governments also are treated as large local governments for these purposes (without regard to population).

Each State or large local government receiving an allocation of QEC bonds may further allocate issuance authority to issuers within such State or large local government. However, any allocations to issuers within the State or large local government must be made in a manner that results in not less than 70 percent of the allocation of QEC bonds to such State or large local government being used to designate bonds that are not private activity bonds (i.e., the bond cannot meet the private business tests or the private loan test of section 141); for purposes of applying this rule, any bond used for the purpose of providing grants, loans or other repayment mechanisms for capital expenditures to implement green community programs is not treated as a private activity bond.

In the case of QEC bonds issued as private activity bonds, 100 percent of the available project proceeds must be used for capital expenditures[76]. In addition, QEC bonds only may be issued by Indian tribal governments to the extent such bonds are issued for purposes that satisfy the present law requirements for tax-exempt bonds issued by Indian tribal governments (i.e., essential governmental functions and certain manufacturing purposes).[77]

One-hundred percent of the available project proceeds with respect to an issuance of QEC bonds must be used within the three-year period that begins on the date of issuance.[78] Available project proceeds are proceeds from the sale of the issue less issuance costs (not to exceed two percent) and any investment earnings on such sale proceeds. To the extent that less than 100 percent of the available project proceeds with respect to an issuance of QEC bonds are used to finance qualified conservation purposes during the three-year spending period, such bonds will continue to qualify as QEC bonds if unspent proceeds are used within 90 days from the end of such three-year period to redeem such bonds. The three-year spending period may be extended by the Secretary upon the issuer's request if the issuer establishes that the failure to satisfy the three-year requirement is due to reasonable cause and the qualified expenditures will continue to proceed with due diligence.

QEC bonds generally are subject to the arbitrage requirements of section 148.[79] However, available project proceeds invested during the three-year spending period are not subject to the arbitrage restrictions (i.e., yield restriction and rebate requirements). In addition, amounts invested in a reserve fund are not subject to the arbitrage restrictions to the extent: (1) such fund is funded at a rate not more rapid than equal annual installments; (2) such fund is funded in a manner reasonably expected to result in an amount not greater than an amount necessary to repay the issue; and (3) the yield on such fund is not greater than the average annual interest rate of tax-exempt obligations having a term of 10 years or more that are issued during the month the QEC bonds are issued.

The maturity of QEC bonds is the term that the Secretary estimates will result in the present value of the obligation to repay the principal on such bonds being equal to 50 percent of the face amount of such bonds, using as a discount rate the average annual interest rate of tax- exempt obligations having a term of 10 years or more that are issued during the month the QEC bonds are issued.[80]

As with other tax credit bonds, the taxpayer holding QEC bonds on a credit allowance date is entitled to a tax credit.[81] The credit rate on the bonds is set by the Secretary at a rate that is 70 percent of the rate that would permit issuance of such bonds without discount and interest cost to the issuer.[82] The Secretary determines credit rates for tax credit bonds based on general assumptions about credit quality of the class of potential eligible issuers and such other factors as the Secretary deems appropriate. The Secretary may determine credit rates based on general credit market yield indexes and credit ratings.[83] The amount of the tax credit is determined by multiplying the bond's credit rate by the face amount on the holder's bond. The credit accrues quarterly, is includible in gross income (as if it were an interest payment on the bond), and can be claimed against regular income tax liability and alternative minimum tax liability. Unused credits may be carried forward to succeeding taxable years. In addition, credits may be separated from the ownership of the underlying bond similar to how interest coupons can be stripped for interest-bearing bonds.

Issuers of QEC bonds are required to certify that the financial disclosure requirements that applicable State and local law requirements governing conflicts of interest are satisfied with respect to such issue, as well as any other additional conflict of interest rules prescribed by the Secretary with respect to any Federal, State, or local government official directly involved with the issuance of QEC bonds.

The following table provides the allocation to States of the National Bond Volume cap for qualified energy conservation bonds.

Table 8. Allocations to States of the National Bond Volume Cap for Qualified Energy Conservations Bonds

State or Territory	QEC Bond Allocations (in dollars)
Alabama	48,364,000
Alaska	7,120,000
Arizona	67,436,000
Arkansas	29,623,000
California	381,329,000
Colorado	51,244,000
Connecticut	36,323,000
Delaware	9,058,000
District of Columbia	6,140,000
Florida	190,146,000
Georgia	100,484,000
Hawaii	13,364,000
Idaho	15,809,000
Illinois	133,846,000

Table 8. (Continued)

State or Territory	QEC Bond Allocations (in dollars)
Indiana	66,155,000
Iowa	31,150,000
Kansas	29,070,000
Kentucky	44,291,000
Louisiana	45,759,000
Maine	13,657,000
Maryland	58,445,000
Massachusetts	67,413,000
Michigan	103,780,000
Minnesota	54,159,000
Mississippi	30,486,000
Missouri	61,329,000
Montana	10,037,000
Nebraska	18,502,000
Nevada	26,975,000
New Hampshire	13,651,000
New Jersey	90,078,000
New Mexico	20,587,000
New York	202,200,000
North Carolina	95,677,000
North Dakota	6,655,000
Ohio	119,160,000
Oklahoma	37,787,000
Oregon	39,320,000
Pennsylvania	129,144,000
Rhode Island	10,901,000
South Carolina	46,475,000
South Dakota	8,343,000
Tennessee	64,476,000
Texas	252,378,000
Utah	28,389,000
Vermont	6,445,000
Virginia	80,600,000
Washington	67,944,000
West Virginia	18,824,000
Wisconsin	58,387,000
Wyoming	5,526,000
American Samoa	673,000
Guam	1,826,000
Northern Marianas	899,000
Puerto Rico	41,021,000
US Virgin Islands	1,140,000
Total Allocation	3,200,000,000

7. Energy efficient commercial buildings deduction (sec. 179D)

In general

Code section 179D provides a deduction equal to energy-efficient commercial building property expenditures made by the taxpayer. Energy-efficient commercial building property expenditures are defined as property (1) which is installed on or in any building located in the United States that is within the scope of Standard 90.1-2001 of the American Society of Heating, Refrigerating, and Air Conditioning Engineers and the Illuminating Engineering Society of North America ("ASHRAE/IESNA"), (2) which is installed as part of (i) the interior lighting systems, (ii) the heating, cooling, ventilation, and hot water systems, or (iii) the building envelope, and (3) which is certified as being installed as part of a plan designed to reduce the total annual energy and power costs with respect to the interior lighting systems, heating, cooling, ventilation, and hot water systems of the building by 50 percent or more in comparison to a reference building which meets the minimum requirements of Standard 90.1-2001 (as in effect on April 2, 2003). The deduction is limited to an amount equal to $1.80 per square foot of the property for which such expenditures are made. The deduction is allowed in the year in which the property is placed in service.

Certain certification requirements must be met in order to qualify for the deduction. The Secretary, in consultation with the Secretary of Energy, will promulgate regulations that describe methods of calculating and verifying energy and power costs using qualified computer software based on the provisions of the 2005 California Nonresidential Alternative Calculation Method Approval Manual or, in the case of residential property, the 2005 California Residential Alternative Calculation Method Approval Manual.

The Secretary is granted authority to prescribe procedures for the inspection and testing for compliance of buildings that are comparable, given the difference between commercial and residential buildings, to the requirements in the Mortgage Industry National Accreditation Procedures for Home Energy Rating Systems.[84] Individuals qualified to determine compliance shall only be those recognized by one or more organizations certified by the Secretary for such purposes.

For energy-efficient commercial building property expenditures made by a public entity, such as public schools, the Secretary shall promulgate regulations that allow the deduction to be allocated to the person primarily responsible for designing the property in lieu of the public entity.

If a deduction is allowed under this section, the basis of the property shall be reduced by the amount of the deduction.

The deduction is effective for property placed in service prior to January 1, 2014.

Partial allowance of deduction

In the case of a building that does not meet the overall building requirement of a 50-percent energy savings, a partial deduction is allowed with respect to each separate building system that comprises energy efficient property and which is certified by a qualified professional as meeting or exceeding the applicable system-specific savings targets established by the Secretary of the Treasury. The applicable system-specific savings targets to be established by the Secretary are those that would result in a total annual energy savings with respect to the whole building of 50 percent, if each of the separate systems met the system specific target. The separate building systems are (1) the interior lighting system, (2) the heating, cooling, ventilation and hot water systems, and (3) the building envelope. The maximum allowable deduction is $0.60 per square foot for each separate system.

Interim rules for lighting systems

In general, in the case of system-specific partial deductions, no deduction is allowed until the Secretary establishes system-specific targets.[85] However, in the case of lighting system retrofits, until such time as the Secretary issues final regulations, the system-specific energy savings target for the lighting system is deemed to be met by a reduction in Lighting Power Density of 40 percent (50 percent in the case of a warehouse) of the minimum requirements in Table 9.3.1.1 or Table 9.3.1.2 of ASHRAE/IESNA Standard 90.1-2001. Also, in the case of a lighting system that reduces lighting power density by 25 percent, a partial deduction of 30 cents per square foot is allowed. A pro-rated partial deduction is allowed in the case of a lighting system that reduces lighting power density between 25 percent and 40 percent. Certain lighting level and lighting control requirements must also be met in order to qualify for the partial lighting deductions under the interim rule.

8. Energy conservation subsidies provided by public utilities (sec. 136)

An exclusion from gross income is provided for the value of any subsidy provided by a public utility to a customer for the purchase or installation of any energy conservation measure, meaning any installation or modification primarily designed to reduce consumption of electricity or natural gas or to improve the management of energy demand with respect to a dwelling unit.

No deduction or credit is allowed for any expenditure to the extent of the exclusion taken for any subsidy received, and the adjusted basis of the property is reduced by the amount excluded.

A "public utility" means a person engaged in the sale of electricity or natural gas to residential, commercial, or industrial customers for use by such customers, and such term includes the Federal Government, a State or local government, or any political subdivision or instrumentality thereof. The exclusion does not apply with respect to any payment to or from a qualified cogeneration facility or qualifying small power production facility pursuant to section 210 of the Public Utility Regulatory Policy Act of 1978.

C. Provisions Relating to Fossil Fuel Extraction

1. Credit for enhanced oil recovery costs (sec. 43)

Taxpayers may claim a credit equal to 15 percent of enhanced oil recovery ("EOR") costs.[86] Qualified EOR costs include the following costs associated with an EOR project. An EOR project is generally a project that involves the use of one or more tertiary recovery methods (as defined in section 1 93(b)(3)) to increase the amount of recoverable domestic crude oil. Qualified costs include (1) amounts paid for depreciable tangible property; (2) intangible drilling and development expenses; (3) tertiary injectant expenses; and (4) construction costs for certain Alaskan natural gas treatment facilities.

The EOR credit is ratably reduced over a $6 phase-out range when the reference price for domestic crude oil exceeds $28 per barrel (adjusted for inflation after 1991). The reference price is determined based on the annual average price of domestic crude oil for the calendar year preceding the calendar year in which the taxable year begins.[87] The EOR credit is currently phased-out.

Taxpayers claiming the EOR credit must reduce by the amount of the credit any otherwise allowable deductions associated with EOR costs. In addition, to the extent a property's basis would otherwise be increased by any EOR costs, such basis is reduced by the amount of the EOR credit.

2. Credit for the production of Indian coal (sec. 45)

A credit is available for the production of Indian coal sold to an unrelated third party from a qualified facility for a seven-year period beginning January 1, 2006, and ending December 31, 2012. The amount of the credit for Indian coal is $1.50 per ton for the first four years of the seven-year period and $2.00 per ton for the last three years of the seven-year period. Beginning in calendar years after 2006, the credit amounts are indexed annually for inflation using 2005 as the base year.

A qualified Indian coal facility is a facility placed in service before January 1, 2009, that produces coal from reserves that on June 14, 2005, were owned by a Federally recognized tribe of Indians or were held in trust by the United States for a tribe or its members.

The credit is a component of the general business credit,[88] allowing excess credits to be carried back one year and forward up to 20 years. The credit is also subject to the alternative minimum tax.

3. Marginal wells credit (sec. 45I)

The Code provides a $3 per-barrel credit (adjusted for inflation) for the production of crude oil and a $0.50-per-1,000-cubic-feet credit (also adjusted for inflation) for the production of qualified natural gas. In both cases, the credit is available only for domestic production from a "qualified marginal well."

A qualified marginal well is defined as a domestic well: (1) production from which is treated as marginal production for purposes of the Code percentage depletion rules; or (2) that during the taxable year had average daily production of not more than 25 barrel equivalents and produces water at a rate of not less than 95 percent of total well effluent. The maximum amount of production on which credit could be claimed is 1,095 barrels or barrel equivalents.

The credit is not available to production occurring if the reference price of oil exceeds $18 ($2.00 for natural gas). The credit is reduced proportionately as for reference prices between $15 and $18 ($1.67 and $2.00 for natural gas). Currently the credit is phased out completely.

In the case of production from a qualified marginal well which is eligible for the credit allowed under section 45K for the taxable year, no marginal well credit is allowable unless the taxpayer elects not to claim the credit under section 45K with respect to the well. The section 45K credit is currently expired with respect to qualified natural gas and oil production. The credit is treated as a general business credit. Unused credits can be carried back for up to five years rather than the generally applicable carryback period of one year.

4. Recovery of intangible drilling and development costs (secs. 59, 263, and 291)

The Code provides special rules for the treatment of intangible drilling and development costs ("IDCs"). Under these special rules, an operator or working interest owner[89] that pays or incurs IDCs in the development of an oil or gas property located in the United States may elect either to expense or capitalize those costs.[90]

IDCs include all expenditures made by an operator for wages, fuel, repairs, hauling, supplies, etc., incident to and necessary for the drilling of wells and the preparation of wells for the production of oil and gas. In addition, IDCs include the cost to operators of any drilling or development work done by contractors under any form of contract, including a turnkey contract. Such work includes labor, fuel, repairs, hauling, and supplies which are used (1) in the drilling, shooting, and cleaning of wells; (2) in the clearing of ground, draining, road making, surveying, and geological works as necessary in preparation for the drilling of wells; and (3) in the construction of such derricks, tanks, pipelines, and other physical structures as are necessary for the drilling of wells and the preparation of wells for the production of oil and gas. Generally, IDCs do not include expenses for items that have a salvage value (such as pipes and casings) or items that are part of the acquisition price of an interest in the property.[91] They also do not include the cost to operators (1) payable only out of production or gross or net proceeds from production, if the amounts are depletable income to the recipient, and (2) amounts properly allocable to the cost of depreciable property.

If an election to expense IDCs is made, the taxpayer deducts the amount of the IDCs as an expense in the taxable year the cost is paid or incurred. Generally, if IDCs are not expensed, but are capitalized, they may be recovered through depletion or depreciation, as appropriate. In the case of a nonproductive well ("dry hole"), IDCs may be deducted at the election of the operator.[92] For an integrated oil company that has elected to expense IDCs, 30 percent of the IDCs on productive wells must be capitalized and amortized over a 60-month period.[93]

Notwithstanding the fact that a taxpayer has made the election to deduct IDCs, the Code provides an additional election under which the taxpayer is allowed to capitalize and amortize certain IDCs over a 60-month period beginning with the month the expenditure was paid or incurred.[94] This election applies on an expenditure-by-expenditure basis; that is, for any particular taxable year, a taxpayer may deduct some portion of its IDCs and capitalize the rest under this provision. The election allows a taxpayer to reduce or eliminate the IDC adjustments or preferences under the alternative minimum tax.

The election to deduct IDCs applies only to those IDCs associated with domestic properties.[95] For this purpose, the United States includes certain wells drilled offshore.[96]

Pursuant to a special exception, the uniform capitalization rules do not apply to IDCs incurred with respect to oil or gas wells that are otherwise deductible under the Code.[97]

5. Deduction for qualified tertiary injectants (sec. 193)

Taxpayers engaged in petroleum extraction activities may generally deduct qualified tertiary injectant expenses used while applying a tertiary recovery method, including carbon dioxide augmented waterflooding and immiscible carbon dioxide displacement.[98] The deduction is available even if such costs are otherwise subject to capitalization. The deduction is permitted for the later of--(1) the tax year in which the injectant is injected or (2) the tax year in which the expenses are paid or incurred.[99] No deduction is permitted for expenditures for which a taxpayer has elected to deduct such costs under section 263(c) (intangible drilling costs) or if a deduction is allowed for such amounts under any other income tax provision. [100]

A "qualified tertiary injectant expense" is defined as any cost paid or incurred for any tertiary injectant (other than a recoverable hydrocarbon injectant) which is used as part of a tertiary recovery method.[101] The cost of a recoverable hydrocarbon injectant (which includes natural gas, crude oil and any other injectant with more than an insignificant amount of natural gas or crude oil) is not a qualified tertiary injectant expense unless the amount of the recoverable hydrocarbon injectant in the qualified tertiary injectant is insignificant.[102]

6. Arbitrage exception for prepayments of natural gas (sec. 148)

Arbitrage restrictions

Interest on bonds issued by States or local governments to finance activities carried out or paid for by those entities generally is exempt from income tax. Restrictions are imposed on the ability of States or local governments to invest the proceeds of these bonds for profit (the "arbitrage restrictions").[103] One such restriction limits the use of bond proceeds to acquire "investment-type property." The term investment-type property includes the acquisition of property in a transaction involving a prepayment if a principal purpose of the prepayment is to receive an investment return from the time the prepayment is made until the time payment otherwise would be made. A prepayment can produce prohibited arbitrage profits when the discount received for prepaying the costs exceeds the yield on the tax-exempt bonds. In general, prohibited prepayments include all prepayments that are not customary in an industry by both beneficiaries of tax-exempt bonds and other persons using taxable financing for the same transaction.

Arbitrage exceptions for certain prepayments

On August 4, 2003, the Treasury Department issued final regulations deeming to be customary, and not in violation of the arbitrage rules, certain prepayments for natural gas and electricity.[104] Generally, a qualified prepayment under the regulations requires that 90 percent of the natural gas or electricity purchased with the prepayment be used for a qualifying use. Generally, natural gas is used for a qualifying use if it is to be (1) furnished to retail gas customers of the issuing municipal utility who are located in the natural gas service area of the issuing municipal utility, however, gas used to produce electricity for sale is not included under this provision; (2) used by the issuing municipal utility to produce electricity that will be furnished to retail electric service area customers of the issuing utility; (3) used by the issuing municipal utility to produce electricity that will be sold to a utility owned by a governmental person and furnished to the service area retail electric customers of the purchaser; (4) sold to a utility that is owned by a governmental person if the requirements of (1), (2) or (3) are satisfied by the purchasing utility (treating the purchaser as the issuing utility); or (5) used to fuel the pipeline transportation of the prepaid gas supply. Electricity is used for a qualifying use if it is to be (1) furnished to retail service area electric customers of the issuing municipal utility or (2) sold to a municipal utility and furnished to retail electric customers of the purchaser who are located in the electricity service area of the purchaser.

The Energy Policy Act of 2005 (the "Act") created a statutory exception to the general rule that tax-exempt bond-financed prepayments violate the arbitrage restrictions. Under the

Act, the term "investment type property" does not include a prepayment under a qualified natural gas supply contract.

A contract is a qualified natural gas contract if the volume of natural gas secured for any year covered by the prepayment does not exceed the sum of (1) the average annual natural gas purchased (other than for resale) by customers of the utility within the service area of the utility ("retail natural gas consumption") during the testing period, and (2) the amount of natural gas that is needed to fuel transportation of the natural gas to the governmental utility. The testing period is the 5-calendar-year period immediately preceding the calendar year in which the bonds are issued. A retail customer is one who does not purchase natural gas for resale. Natural gas used to generate electricity by a utility owned by a governmental unit is counted as retail natural gas consumption if the electricity was sold to retail customers within the service area of the governmental electric utility.

Adjustments

The volume of gas permitted by the general rule is reduced by natural gas otherwise available on the date of issuance. Specifically, the amount of natural gas permitted to be acquired under a qualified natural gas contract for any period is to be reduced by the applicable share of natural gas held by the utility on the date of issuance of the bonds and natural gas that the utility has a right to acquire for the prepayment period (determined as of the date of issuance). For purposes of the preceding sentence, "applicable share" means, with respect to any period, the natural gas allocable to such period if the gas were allocated ratably over the period to which the prepayment relates.

For purposes of the safe harbor, if after the close of the testing period and before the issue date of the bonds (1) the government utility enters into a contract to supply natural gas (other than for resale) for a commercial person for use at a property within the service area of such utility and (2) the gas consumption for such property was not included in the testing period or the ratable amount of natural gas to be supplied under the contract is significantly greater than the ratable amount of gas supplied to such property during the testing period, then the amount of gas permitted to be purchased may be increased to accommodate the contract.

The calculation of average annual retail natural gas consumption for purposes of the safe harbor, however, is not to exceed the annual amount of natural gas reasonably expected to be purchased (other than for resale) by persons who are located within the service area of such utility and who, as of the date of issuance of the issue, are customers of such utility.

Intentional acts

The safe harbor does not apply if the utility engages in intentional acts to render (1) the volume of natural gas covered by the prepayment to be in excess of that needed for retail natural gas consumption, and (2) the amount of natural gas that is needed to fuel transportation of the natural gas to the governmental utility.

Definition of service area

Service area is defined as (1) any area throughout which the governmental utility provided (at all times during the testing period) in the case of a natural gas utility, natural gas transmission or distribution services, or in the case of an electric utility, electricity distribution services; (2) limited areas contiguous to such areas; and (3) any area recognized as the service area of the governmental utility under State or Federal law. Contiguous areas are limited to any area within a county contiguous to the area described in item (1) above in which retail customers of the utility are located if such area is not also served by another utility providing the same service.

Ruling request for higher prepayment amounts

Upon written request, the Secretary may allow an issuer to prepay for an amount of gas greater than that allowed by the safe harbor based on objective evidence of growth in gas consumption or population that demonstrates that the amount permitted by the exception is insufficient.

Application to joint action agencies

In a number of States, joint action agencies serve as purchasing agents for their member municipal gas utilities. The provision is intended to allow municipal utilities in a State to participate in such buying arrangements as established under State law, subject to the same limitations that would apply if an individual utility were to purchase gas directly. When acting on behalf of its municipal gas utility members, the total amount of gas that can be purchased by a joint action agency under the provision's exception to the arbitrage rules is the aggregate of what each such member could purchase for itself on a direct basis. Thus, with respect to qualified natural gas supply contracts entered into by joint action agencies for or on behalf of one or more member municipal utilities, the requirements of the safe harbor are tested at the individual municipal utility level based on the amount of gas that would be allocated to such member during any year covered by the contract.

7. Amortization of geological and geophysical expenditures (sec. 167(h))

Geological and geophysical expenditures ("G&G costs") are costs incurred by a taxpayer for the purpose of obtaining and accumulating data that will serve as the basis for the acquisition and retention of mineral properties by taxpayers exploring for minerals. G&G costs incurred by independent producers and smaller integrated oil companies in connection with oil and gas exploration in the United States may generally be amortized over two years.[105]

Major integrated oil companies are required to amortize all G&G costs over seven years for costs paid or incurred after December 19, 2007 (the date of enactment of the Energy Independence and Security Act of 2007, Pub. L. No. 110-140).[106] A major integrated oil company, as defined in section 167(h)(5)(B), is an integrated oil company[107] which has an average daily worldwide production of crude oil of at least 500,000 barrels for the taxable year, had gross receipts in excess of one billion dollars for its last taxable year ending during the calendar year 2005, and generally has an ownership interest in a crude oil refiner of 15 percent or more.

In the case of abandoned property, remaining basis may not be recovered in the year of abandonment of a property, but instead must continue to be amortized over the remaining applicable amortization period.

8. Passive loss rules for working interests in oil and gas property (sec. 469)

Present Law

The passive loss rules limit deductions and credits from passive trade or business activities.[108] Deductions attributable to passive activities, to the extent they exceed income from passive activities, generally may not be deducted against other income. Deductions and credits that are suspended under these rules are carried forward and treated as deductions and credits from passive activities in the next year. The suspended losses from a passive activity are allowed in full when a taxpayer disposes of his entire interest in the passive activity to an unrelated person.

Losses from certain working interests in oil and gas property are not limited under the passive loss rule. Thus, losses and credits from such interests can be used to offset other income of the taxpayer without limitation under the passive loss rule. A passive activity does not include a working interest in any oil or gas property that the taxpayer holds directly or through an entity that does not limit the liability of the taxpayer with respect to the interest. This rule applies without regard to whether the taxpayer materially participates in the activity. If the taxpayer has a loss from a working interest in any oil or gas property that is treated as not from a passive activity, then net income from the property for any succeeding taxable year is treated as income of the taxpayer that is not from a passive activity.

In general, a working interest is an interest with respect to an oil and gas property that is burdened with the cost of development and operation of the property. Rights to overriding royalties, production payments, and the like, do not constitute working interests, because they are not burdened with the responsibility to share expenses of drilling, completing, or operating oil and gas property. Similarly, contract rights to extract or share in oil and gas, or in profits from extraction, without liability to share in the costs of production, do not constitute working interests. Income from such interests generally is considered to be portfolio income.

A working interest generally has characteristics such as responsibility for signing authorizations for expenditures with respect to the activity, receiving periodic drilling and completion reports, receiving periodic reports regarding the amount of oil extracted, possession of voting rights proportionate to the percentage of the working interest possessed by the taxpayer, the right to continue activities if the present operator decides to discontinue operations, a proportionate share of tort liability with respect to the property (e.g., if a well catches fire), and some responsibility to share in further costs with respect to the property in the event that a decision is made to spend more than amounts already contributed.

However, the fact that a taxpayer is entitled to decline, or does decline, to make additional contributions under a buyout, nonparticipation, or similar arrangement, does not prevent such taxpayer's possessing a working interest. In addition, the fact that tort liability may be insured against does not prevent such taxpayer's possessing a working interest.

When the taxpayer's form of ownership limits the liability of the taxpayer, the interest possessed by such taxpayer is not a working interest for purposes of the passive loss provision. Thus, for purposes of the passive loss rules, an interest owned by a limited partnership is not treated as a working interest with regard to any limited partner, and an interest owned by an S corporation is not treated as a working interest with regard to any shareholder. The same result follows with respect to any form of ownership that is substantially equivalent in its effect on liability to a limited partnership interest or interest in an S corporation, even if different in form.

When an interest is not treated as a working interest because the taxpayer's form of ownership limits his liability, the general rules regarding material participation apply to determine whether the interest is treated as in a passive activity. Thus, for example, a limited partner's interest generally is treated as in a passive activity. In the case of a shareholder in an S corporation, the general facts and circumstances test for material participation applies and the working interest exception does not apply, because the form of ownership limits the taxpayer's liability.

In determining whether the taxpayer's form of ownership limits his liability, the rule described in the two prior paragraphs is applied by looking through tiered entities. For example, a general partner in a partnership that owns a limited partnership interest in a partnership that owns a working interest is not treated as owning a working interest.

A special rule applies in any case where, for a prior taxable year, net losses from a working interest in a property were treated by the taxpayer as not from a passive activity. In such a case, any net income realized by the taxpayer from the property (or from any substituted basis property, e.g., property acquired in a sec. 1031 like kind exchange for such property) in a subsequent year also is treated as active. Under this rule, for example, if a taxpayer claims losses for a year with regard to a working interest and then, after the property to which the interest relates begins to generate net income, transfers the interest to an S corporation in which he is a shareholder, or to a partnership in which he has an interest as a limited partner, his interest with regard to the property continues to be treated as not passive.

9. Depletion (secs. 611-613A)

In general

Depletion, like depreciation, is a form of capital cost recovery. In both cases, the taxpayer is allowed a deduction in recognition of the fact that an asset is being expended in order to produce income.[109] Certain costs incurred prior to drilling an oil or gas property or extracting minerals are recovered through the depletion deduction. These include the cost of acquiring the lease or other interest in the property.

Depletion is available to any person having an economic interest in a producing property. An economic interest is possessed in every case in which the taxpayer has acquired by investment any interest in minerals in place, and secures, by any form of legal relationship, income derived from the extraction of the mineral, to which it must look for a return of its capital. Thus, for example, both working interests and royalty interests in an oil- or gas-producing property constitute economic interests, thereby qualifying the interest holders for depletion deductions with respect to the property. A taxpayer who has no capital investment in the mineral deposit, however, does not acquire an economic interest merely by possessing an economic or pecuniary advantage derived from production through a contractual relation.

Two methods of depletion are currently allowable under the Code: (1) the cost depletion method, and (2) the percentage depletion method.[110] Under the cost depletion method, the taxpayer deducts that portion of the adjusted basis of the depletable property which is equal to the ratio of units sold from that property during the taxable year to the number of units remaining as of the end of taxable year plus the number of units sold during the taxable year.

Thus, the amount recovered under cost depletion may never exceed the taxpayer's basis in the property.

Under the percentage depletion method, a percentage, varying from five percent to 22 percent (generally 15 percent for oil and gas properties), of the taxpayer's gross income from a producing property is allowed as a deduction in each taxable year.[111] The amount deducted generally may not exceed 50 percent (100 percent in the case of oil and gas properties) of the net income from the oil and gas property in any year (the "net-income limitation").[112] Additionally, the percentage depletion deduction for all oil and gas properties may not exceed 65 percent of the taxpayer's overall taxable income for the year (determined before such deduction and adjusted for certain loss carrybacks and trust distributions).[113]

The percentage depletion rate for coal and lignite is 10 percent.[114] The percentage depletion rate for oil shale is 15 percent.[115] In the case of iron ore and coal (including lignite), a corporate preference reduces the amount of percentage depletion calculated by 20 percent of the amount of percentage depletion in excess of the adjusted basis of the property at the close of the taxable year (determined without regard to the depletion deduction for the taxable year).[116]

Because percentage depletion, unlike cost depletion, is computed without regard to the taxpayer's basis in the depletable property, cumulative depletion deductions may be greater than the amount expended by the taxpayer to acquire or develop the property. A taxpayer is required to determine the depletion deduction for each property under both the percentage depletion method (if the taxpayer is entitled to use this method) and the cost depletion method. If the cost depletion deduction is larger, the taxpayer must utilize that method for the taxable year in question.[117]

Limitation on oil and gas percentage depletion to independent producers and royalty owners

As stated above, percentage depletion of oil and gas properties generally is not permitted, except for independent producers and royalty owners, certain fixed-price gas contracts, and natural gas from geopressured brine. For purposes of the percentage depletion allowance, an independent producer is any producer that is not a "retailer" or "refiner." A retailer is any person that directly, or through a related person, sells oil or natural gas (or a derivative thereof):

(1) through any retail outlet operated by the taxpayer or related person, or
(2) to any person that is obligated to market or distribute such oil or natural gas (or a derivative thereof) under the name of the taxpayer or the related person, or that has the authority to occupy any retail outlet owned by the taxpayer or a related person.[118]

Bulk sales of crude oil and natural gas to commercial or industrial users, and bulk sales of aviation fuel to the Department of Defense, are not treated as retail sales. Further, if the combined gross receipts of the taxpayer and all related persons from the retail sale of oil, natural gas, or any product derived therefrom do not exceed $5 million for the taxable year, the taxpayer will not be treated as a retailer.

A refiner is any person that directly or through a related person engages in the refining of crude oil in excess of an average daily refinery run of 75,000 barrels during the taxable year.[119]

Percentage depletion for eligible taxpayers is allowed for up to 1,000 barrels of average daily production of domestic crude oil or an equivalent amount of domestic natural gas.[120] For producers of both oil and natural gas, this limitation applies on a combined basis. All production owned by businesses under common control and members of the same family must be aggregated;[121] each group is then treated as one producer in applying the 1,000-barrel limitation.

In addition to independent producers and royalty owners, certain sales of natural gas under a fixed contract in effect on February 1, 1975, and certain natural gas from geopressured brine, are eligible for percentage depletion, at rates of 22 percent and 10 percent, respectively. These exceptions apply without regard to the 1,000-barrel-per-day limitation and regardless of whether the producer is an independent producer or an integrated oil company.

Before enactment of the Omnibus Budget Reconciliation Act of 1990 (the "1990 Act"), if an interest in a proven oil or gas property was transferred (subject to certain exceptions), the production from such interest did not qualify for percentage depletion. The 1990 Act repealed the limitation on claiming percentage depletion on transferred properties effective for property transfers occurring after October 11, 1990.

Percentage depletion on marginal production

The 1990 Act also created a special percentage depletion provision for oil and gas production from so-called marginal properties held by independent producers or royalty owners.[122] Under this provision, the statutory percentage depletion rate is increased (from the general rate of 15 percent) by one percent for each whole dollar that the average price of crude oil for the immediately preceding calendar year is less than $20 per barrel. In no event may the rate of percentage depletion under this provision exceed 25 percent for any taxable year. The increased rate applies for the taxpayer's taxable year that immediately follows a calendar year for which the average crude oil price falls below the $20 floor. To illustrate the application of this provision, the average price of a barrel of crude oil for calendar year 1999 was $15.56. Thus, the percentage depletion rate for production from marginal wells was increased to 19 percent for taxable years beginning in 2000. Since the price of oil currently is above the $20 floor, there is no increase in the statutory depletion rate for marginal production.

The Code defines the term "marginal production" for this purpose as domestic crude oil or domestic natural gas which is produced during any taxable year from a property which (1) is a stripper well property for the calendar year in which the taxable year begins, or (2) is a property substantially all of the production from which during such calendar year is heavy oil (i.e., oil that has a weighted average gravity of 20 degrees API or less, corrected to 60 degrees Fahrenheit).[123] A stripper well property is any oil or gas property that produces a daily average of 15 or fewer equivalent barrels of oil and gas per producing oil or gas well on such property in the calendar year during which the taxpayer's taxable year begins.[124]

The determination of whether a property qualifies as a stripper well property is made separately for each calendar year. The fact that a property is or is not a stripper well property for one year does not affect the determination of the status of that property for a subsequent year. Further, a taxpayer makes the stripper well property determination for each separate property interest (as defined under section 614) held by the taxpayer during a calendar year. The determination is based on the total amount of production from all producing wells that are treated as part of the same property interest of the taxpayer. A property qualifies as a stripper well property for a calendar year only if the wells on such property were producing during that period at their maximum efficient rate of flow.

If a taxpayer's property consists of a partial interest in one or more oil- or gas-producing wells, the determination of whether the property is a stripper well property or a heavy oil property is made with respect to total production from such wells, including the portion of total production attributable to ownership interests other than the taxpayer's interest. If the property satisfies the requirements of a stripper well property, then the benefits of this provision apply with respect to the taxpayer's allocable share of the production from the property. The deduction is allowed for the taxable year that begins during the calendar year in which the property so qualifies.

The allowance for percentage depletion on production from marginal oil and gas properties is subject to the 1,000-barrel-per-day limitation discussed above. Unless a taxpayer elects otherwise, marginal production is given priority over other production for purposes of utilization of that limitation.

D. Provisions Relating to Refining

1. Tax incentives for refining (secs. 45H, 168, 179B, and 179C)

General rule for depreciating refinery assets (sec. 168)

A taxpayer is allowed to recover, through annual depreciation deductions, the cost of certain property used in a trade or business or for the production of income. The amount of the depreciation deduction allowed with respect to tangible property for a taxable year is determined under the modified accelerated cost recovery system ("MACRS"). The applicable recovery period for assets placed in service under the MACRS is based on the "class life of the property." Except where provided specifically by statute, the class lives of assets placed in service after 1986 are generally set forth in Revenue Procedure 87-56.[125] In the Revenue Procedure, petroleum refining assets are assigned a 10-year recovery period and a class life of 16 years. Petroleum refining assets are assets used for distillation, fractionation, and catalytic cracking of crude petroleum into gasoline and its other components.

Temporary election to expense 50 percent of qualified property used in refining liquid fuels (sec. 179C)

Taxpayers may elect to expense 50 percent of the cost of any qualified property used for processing liquid fuel from crude oil or qualified fuels (as defined in section 45K(c)[126] of the Code). The remaining 50 percent is recovered under the otherwise applicable rules. Qualified refinery property includes assets located in the United States that are used in the refining of liquid fuels: (1) with respect to the construction of which a binding construction contract has been entered into before January 1, 2010;[127] (2) which are placed in service before January 1, 2014; (3) which increase the capacity of an existing refinery by at least five percent or increase the percentage of total throughput attributable to qualified fuels such that it equals or exceeds 25 percent; and (4) which meet all applicable environmental laws in effect when the property is placed in service.

Special expensing rule for capital costs incurred by small refiners to comply with EPA sulfur regulations (sec. 179B)

Under present law, a small business refiner may immediately deduct as an expense 75 percent of the costs paid or incurred for purposes of complying with the Highway Diesel Fuel Sulfur Control requirement of the Environmental Protection Agency ("EPA"). A cooperative that qualifies as a small business refiner may elect to pass this deduction through to its owners.

Costs qualifying for the deduction are those costs paid or incurred with respect to any facility of a small business refiner during the period beginning on January 1, 2003 and ending on the earlier of the date that is one year after the date on which the taxpayer must comply with the applicable EPA regulations or December 31, 2009. A small business refiner is a crude oil refiner that has no more than 1,500 individuals engaged in refinery operations on any given day and that had an average daily domestic refinery run or average retained production of not more than 205,000 barrels for the one-year period ending on December 31, 2002.

Credit for small refiners for production of diesel fuel in compliance with EPA sulfur regulations (sec. 45H)

Under present law, a small business refiner (as defined above) may claim a credit of five cents per gallon for each gallon of low sulfur diesel fuel produced during the taxable year. "Low sulfur diesel fuel" is diesel fuel with a sulfur content of 15 parts per million or less.

The total production credit claimed by the taxpayer is limited to 25 percent of the qualified costs incurred to come into compliance with the EPA diesel fuel requirements. The percentage limitation phases down pro rata for refiners that had runs in 2002 exceeding 155,000 barrels but less than 205,000 barrels.

Costs qualifying for the credit are those costs paid or incurred with respect to any facility of a small business refiner during the period beginning on January 1, 2003 and ending on the earlier of the date that is one year after the date on which the taxpayer must comply with the applicable EPA regulations or December 31, 2009.

2. Credit for producing coke or coke gas (sec. 45K)

Coke and coke gas produced in the United States at qualified facilities and sold to unrelated parties are eligible for an income tax credit equal to $3 (generally adjusted for inflation)[128] per Btu oil barrel equivalent (the "coke credit"). The credit is available for coke or coke gas produced during a four-year period beginning on the later of January 1, 2006, or the date the qualified facility was placed in service. For purposes of the coke credit, qualified facilities are facilities placed in service before January 1, 1993, or after June 30, 1998, and before January 1, 2010. Qualified facilities do not include facilities that produce petroleum-based coke or coke gas. The amount of credit-eligible coke produced at any one facility may not exceed an average barrel-of-oil equivalent of 4,000 barrels per day. The coke credit is part of the general business credit.

E. Provisions Relating to Energy Transportation or Transmission

1. Cost recovery of certain energy transportation and transmission property (sec. 168(e)(3))

In general

A taxpayer is allowed to recover, through annual depreciation deductions, the cost of certain property used in a trade or business or for the production of income. The amount of the depreciation deduction allowed with respect to tangible property for a taxable year is determined under the modified accelerated cost recovery system ("MACRS"). Under MACRS, different types of property are generally assigned applicable recovery periods and depreciation methods. The recovery periods applicable to most tangible personal property (generally tangible property other than residential rental property and nonresidential real property) range from three to 25 years. The depreciation methods generally applicable to tangible personal property are the 200- percent and 150-percent declining balance methods, switching to the straight-line method for the taxable year in which the depreciation deduction would be maximized.[129] In general, the recovery periods applicable to real property are 39 years for non-residential real property and 27.5 years for residential rental property. The depreciation method for real property is the straight-line method.

Under MACRS, the full basis of depreciable property is recovered by the taxpayer over the applicable recovery period; there is no need to estimate salvage value. Furthermore, under MACRS, the applicable recovery period need not (and typically does not) correspond to the actual economic life of the asset subject to depreciation. In general, however, MACRS generally provides for longer recovery periods for longer lived assets.

Alaska natural gas pipeline (sec. 168(e)(3)(C)(iii))

The Code provides a statutory seven year recovery period and a class life of 22 years for any Alaska natural gas pipeline. The term "Alaska natural gas pipeline" is defined as any natural gas pipeline system (including the pipe, trunk lines, related equipment, and appurtenances used to carry natural gas, but not any gas processing plant) located in the State of Alaska that has a capacity of more than 500 billion Btu of natural gas per day and is placed in service after December 31, 2013.[130] A taxpayer who places an otherwise qualifying system in service before January 1, 2014 may elect to treat the system as placed in service on January 1, 2014, thus qualifying for the seven-year recovery period. Absent such an election, the system is subject to the prior law recovery period of 15 years.

Natural gas distribution lines (sec. 168(e)(3)(E)(viii))

The Code provides a statutory 15-year recovery period and a class life of 35 years for natural gas distribution lines. This provision is effective for property, the original use of which begins with the taxpayer after April 11, 2005, and which is placed in service after April 11, 2005 and before January 1, 2011. The provision does not apply to property subject to a binding contract on or before April 11, 2005.[131]

Natural gas gathering lines (sec. 168(e)(3)(C)(iv))

The Code provides a statutory seven-year recovery period and a class life of 14 years for natural gas gathering lines. In addition, no adjustment will be made to the allowable amount of depreciation with respect to this property for purposes of computing a taxpayer's alternative minimum taxable income. A natural gas gathering line is defined to include any pipe, equipment, and appurtenance that is (1) determined to be a gathering line by the Federal Energy Regulatory Commission, or (2) used to deliver natural gas from the wellhead or a common point to the point at which such gas first reaches (a) a gas processing plant, (b) an interconnection with an interstate transmission line, (c) an interconnection with an intrastate transmission line, or (d) a direct interconnection with a local distribution company, a gas storage facility, or an industrial consumer.[132]

Electricity Transmission Property (sec. 168(e)(3)(E)(vii))

Generally, assets used in the transmission and distribution of electricity for sale and related land improvements are assigned a 20-year recovery period and a class life of 30 years. However, the Code provides a statutory 15-year recovery period and a class life of 30 years for certain assets used in the transmission of electricity for sale and related land improvements. Such assets include section 1245 property[133] used in the transmission at 69 or more kilovolts of electricity for sale, the original use of which commences with the taxpayer after April 11, 2005.

2. Five-Year carryback of net operating losses for certain electric utility companies (sec. 172(b)(1)(I))

In general

A net operating loss ("NOL") is, generally, the amount by which a taxpayer's allowable deductions exceed the taxpayer's gross income. A carryback of an NOL generally results in the refund of Federal income tax for the carryback year. A carryover of an NOL reduces Federal income tax for the carryover year. In general, an NOL may be carried back two years and carried over 20 years to offset taxable income in such years.[134] Under present-law ordering rules, NOLs generally are first applied to the earliest of the taxable years to which the loss may be carried.

Certain electric utility company NOLs

The Code provides an election for certain electric utility companies to extend the carryback period to five years for a portion of NOLs arising in 2003, 2004, and 2005 ("loss years"). The election may be made during any taxable year ending after December 31, 2005, and before January 1, 2009 ("election years"). An electing taxpayer must specify to which loss year the election applies.

The portion of the loss year NOL to which the election may apply is limited to 20 percent of the amount of the taxpayer's qualifying investment in the taxable year prior to the year in which the election is made (the "qualifying investment limitation"), with any remaining portion of the loss year NOL subject to the general NOL carryover rules. Only one election may be made in any election year, and elections may not be made for more than one election year beginning in the same calendar year. Thus, for example, a taxpayer with two short taxable years beginning in calendar year 2006 is eligible to make an election under this provision in only one of those two short taxable years. Once an election has been made with respect to a loss year, no subsequent election is available with respect to that loss year.

For purposes of calculating interest on overpayments, any overpayment resulting from a five-year NOL carryback elected under this provision is deemed not to have been made prior to the filing date for the taxable year in which the election is made. The statute of limitations for refund claims, and that for assessment of deficiencies, are also extended.

An election under this provision is made in such manner as the Secretary may prescribe. However, Congress expects that the filing of a refund claim will be considered sufficient for making the election, provided that the taxpayer attaches to the refund claim a statement specifying the election year, the loss year, and the amount of qualifying investment in electric transmission property and pollution control facilities in the preceding taxable year.

Under the provision, an investment in electric transmission property qualifies if it is a capital expenditure made by the taxpayer which is attributable to electric transmission property used by the taxpayer in the transmission at 69 or more kilovolts of electricity for sale. An investment in pollution control equipment qualifies if it is a capital expenditure, made by an electric utility company (as defined in the Public Utility Holding Company Act as in effect on the day before the date of enactment of the provision), which is attributable to a facility which will qualify as a certified pollution control facility, generally as defined under section 169(d)(1) but without regard to the requirements therein that the facility be new or that it be used in connection with a plant or other property in operation before January 1, 1976.

There is no requirement that the transmission property or pollution control facilities be placed in service in the year in which the capital expenditures are incurred. However, it is intended that qualifying investment under the provision includes only capital expenditures to which the taxpayer is committed and with respect to property which the taxpayer intends to ultimately place in service in the taxpayer's trade or business. Under the provision, capital expenditures which, at the taxpayer's option, are refundable or subject to material modification in a manner which would not meet the requirements of the provision, may not be taken into account. For example, if a taxpayer makes a cash deposit with respect to a contract for the purchase of electric transmission property, and the contract contains an option (or there is otherwise an understanding) under which the taxpayer may subsequently apply the deposit to the purchase of equipment other than electric transmission property, the deposit is not included in the taxpayer's qualifying investment. This rule is intended as an anti-abuse rule and should be interpreted to prevent a taxpayer from taking into account capital expenditures to which the taxpayer is not permanently committed.

3. Deferral of gains from the sale of electric transmission property (sec. 451(i))

Generally, a taxpayer recognizes gain to the extent the sales price (and any other consideration received) exceeds the seller's basis in the property. The recognized gain is subject to current income tax unless the gain is deferred or not recognized under a special tax provision. With regard to the disposition of certain electric transmission property, taxpayers may elect to recognize gain from qualified electric transmission transactions ratably over an eight-year period if the amount realized from such sale is used to purchase exempt utility property within four years after the close of the taxable year in which the transaction takes place.[135] If the amount realized exceeds the amount used to purchase reinvestment property, any realized gain shall be recognized to the extent of such excess in the year of the qualifying electric transmission transaction, with any remaining realized gain recognized ratably over the eight-year period.

A qualifying electric transmission transaction is the sale or other disposition of property used by the taxpayer in the trade or business of providing electric transmission services, or an ownership interest in such an entity, to an independent transmission company prior to January 1, 2010. In general, an independent transmission company is defined as: (1) an independent transmission provider[136] approved by the Federal Energy Regulatory Commission ("FERC"); (2) a person (i) who the FERC determines under section 203 of the Federal Power Act (or by declaratory order) is not a "market participant" and (ii) whose transmission facilities are placed under the operational control of a FERC-approved independent transmission provider before the close of the period specified in such authorization, but not later than December 31, 2007; or (3) in the case of facilities subject to the jurisdiction of the Public Utility Commission of Texas, (i) a person which is approved by that Commission as consistent with Texas State law regarding an independent transmission organization, or (ii) a political subdivision, or affiliate thereof, whose transmission facilities are under the operational control of an organization described in (i)

Exempt utility property is defined as: (1) property used in the trade or business of generating, transmitting, distributing, or selling electricity or producing, transmitting, distributing, or selling natural gas, or (2) stock in a controlled corporation whose principal trade or business consists of the activities described in (1).

If the taxpayer is a member of an affiliated group of corporations filing a consolidated return, the reinvestment property may be purchased by any member of the affiliated group (in lieu of the taxpayer).

F. Provisions Relating to Coal and Nuclear Energy

1. Credit for the production of refined coal (sec. 45)

In general

A credit is available for refined coal. In general, refined coal is a fuel produced from coal that is (1) used to produce steam or (2) used to produce steel industry fuel.

Refined coal used to produce steam

An income tax credit is allowed for the production at qualified facilities of certain refined coal sold to an unrelated person. The amount of the refined coal credit is $4.375 per ton (indexed for inflation after 1992 and equaling $6.06 1 per ton for 2008). A taxpayer may generally claim the credit during the 10-year period commencing with the date the qualified facility is placed in service.

A qualifying refined coal facility is a facility producing refined coal that is placed in service after October 22, 2004, and before January 1, 2010. Refined coal is a qualifying liquid, gaseous, or solid synthetic fuel produced from coal (including lignite) or high-carbon fly ash, including such fuel used as a feedstock. A qualifying fuel is a fuel that, when burned, emits 20 percent less nitrogen oxides and either sulfur dioxide or mercury than the burning of feedstock coal or comparable coal predominantly available in the marketplace as of January 1, 2003, but only if the fuel sells at prices at least 50 percent greater than the prices of the feedstock coal or comparable coal. In addition, to be qualified refined coal, the taxpayer must sell the fuel with the reasonable expectation that it will be used for the primary purpose of producing steam.

The refined coal credit is reduced over an $8.75 phase-out range as the reference price of the fuel used as feedstock for the refined coal exceeds an amount equal to 1.7 times the reference price for such fuel in 2002 (adjusted for inflation). The amount of the credit a taxpayer may claim is reduced by reason of grants, tax-exempt bonds, subsidized energy financing, and other credits, but the reduction cannot exceed 50 percent of the otherwise allowable credit.

The credit is a component of the general business credit,[137] allowing excess credits to be carried back one year and forward up to 20 years. The credit is also subject to the alternative minimum tax.

Refined coal facilities used to produce steam that are placed in service after 2008

For refined coal facilities placed in service after 2008, the requirement that the qualified refined coal fuel sell at a price at least 50 percent greater than the price of the feedstock coal does not apply. However, to be credit-eligible, refined coal produced by such facilities must reduce by 40 percent (not 20 percent) the amount by which refined coal must reduce, when burned, emissions of either sulfur dioxide or mercury compared to the emissions released by the feedstock coal or comparable coal predominantly available in the marketplace as of January 1, 2003.

Refined coal that is steel industry fuel

Each barrel-of-oil equivalent (defined as 5.8 million British thermal units) of steel industry fuel produced at a qualified facility during the credit period receives a $2 credit (adjusted for inflation using 1992 as the base year and equal to $2.77 for 2008). A qualified facility is any facility capable of producing steel industry fuel (or any modification to a facility making it so capable) that is placed in service before January 1, 2010. For facilities capable of producing steel industry fuel on or before October 1, 2008, the credit is available for fuel produced and sold on or after such date and before January 1, 2010. For facilities placed in service or modified to produce steel industry fuel after October 1, 2008, the credit period begins on the placed-in-service or modification date and ends one year after such date or December 31, 2009, whichever is later.

Steel industry fuel is defined as a fuel produced through a process of liquefying coal waste sludge, distributing the liquefied product on coal, and using the resulting mixture as a feedstock for the manufacture of coke. Coal waste sludge includes tar decanter sludge and related byproducts of the coking process.

2. Advanced nuclear power production credit (sec. 45J)

Taxpayers producing electricity at a qualifying advanced nuclear power facility may claim a credit equal to 1.8 cents per kilowatt-hour of electricity produced for the eight-year period starting when the facility is placed in service.[138] The aggregate amount of credit that a taxpayer may claim in any year during the eight-year period is subject to limitation based on allocated capacity and an annual limitation as described below.

A qualifying advanced nuclear facility is an advanced nuclear facility for which the taxpayer has received an allocation of megawatt capacity from the Secretary and is placed in service before January 1, 2021. The taxpayer may only claim credit for production of electricity equal to the ratio of the allocated capacity that the taxpayer receives from the Secretary to the rated nameplate capacity of the taxpayer's facility. For example, if the taxpayer receives an allocation of 750 megawatts of capacity from the Secretary and the taxpayer's facility has a rated nameplate capacity of 1,000 megawatts, then the taxpayer may claim three-quarters of the otherwise allowable credit, or 1.35 cents per kilowatt-hour, for each kilowatt-hour of electricity produced at the facility (subject to the annual limitation described below). The Secretary may allocate a total of up to 6,000 megawatts of capacity.

A taxpayer operating a qualified facility may claim no more than \$125 million in tax credits per 1,000 megawatts of allocated capacity in any one year of the eight-year credit period. If the taxpayer operates a 1,350 megawatt rated nameplate capacity system and has received an allocation from the Secretary for 1,350 megawatts of capacity eligible for the credit, the taxpayer's annual limitation on credits that may be claimed is equal to 1.35 times \$125 million, or \$168.75 million. If the taxpayer operates a facility with a nameplate rated capacity of 1,350 megawatts, but has received an allocation from the Secretary for 750 megawatts of credit eligible capacity, then the two limitations apply such that the taxpayer may claim a credit equal to 1 cent per kilowatt-hour of electricity produced (as described above) subject to an annual credit limitation of \$93.75 million in credits (three-quarters of \$125 million).

An advanced nuclear facility is any nuclear facility for the production of electricity, the reactor design for which was approved after 1993 by the Nuclear Regulatory Commission. For this purpose, a qualifying advanced nuclear facility does not include any facility for which a substantially similar design for a facility of comparable capacity was approved before 1994.

In addition, the credit allowable to the taxpayer is reduced by reason of grants, tax-exempt bonds, subsidized energy financing, and other credits, but such reduction cannot exceed 50 percent of the otherwise allowable credit. The credit is treated as part of the general business credit.

3. Credit for carbon dioxide sequestration (sec. 45Q)

A credit of \$10 per metric ton is available for qualified carbon dioxide that is captured by the taxpayer at a qualified facility, used by such taxpayer as a tertiary injectant (including carbon dioxide augmented waterflooding and immiscible carbon dioxide displacement) in a qualified enhanced oil or natural gas recovery project and disposed of by such taxpayer in secure geological storage.[139] In addition, a credit of \$20 per metric ton is available for qualified carbon dioxide captured by a taxpayer at a qualified facility and disposed of by such taxpayer in secure geological storage without being used as a tertiary injectant. Both credit amounts are adjusted for inflation after 2009.

Secure geological storage includes storage at deep saline formations, oil and gas reservoirs, and unminable coal seams. The Secretary, in consultation with the Administrator of the Environmental Protection Agency the Secretary of Energy, and the Secretary of the Interior, is required to establish regulations for determining adequate security measures for the secure geological storage of carbon dioxide such that the carbon dioxide does not escape into the atmosphere.

Qualified carbon dioxide is defined as carbon dioxide captured from an industrial source that (1) would otherwise be released into the atmosphere as an industrial emission of greenhouse gas, and (2) is measured at the source of capture and verified at the point or points of injection. Qualified carbon dioxide includes the initial deposit of captured carbon dioxide used as a tertiary injectant but does not include carbon dioxide that is recaptured, recycled, and re-injected as part of an enhanced oil or natural gas recovery project process. A qualified enhanced oil or natural gas recovery project is a project that would otherwise meet the definition of an enhanced oil recovery project under section 43, if natural gas projects were included within that definition.

A qualified facility means any industrial facility (1) which is owned by the taxpayer, (2) at which carbon capture equipment is placed in service, and (3) which captures not less than 500,000 metric tons of carbon dioxide during the taxable year. The credit applies only with respect to qualified carbon dioxide captured and sequestered or injected in the United States[140] or one of its possessions.[141]

Except as provided in regulations, credits are attributable to the person that captures and physically or contractually ensures the disposal, or use as a tertiary injectant, of the qualified carbon dioxide. Credits are subject to recapture, as provided by regulation, with respect to any qualified carbon dioxide that ceases to be recaptured, disposed of, or used as a tertiary injectant in a manner consistent with the rules of the provision.

The credit is part of the general business credit. The credit sunsets at the end of the calendar year in which the Secretary, in consultation with the Administrator of the Environmental Protection Agency, certifies that 75 million metric tons of qualified carbon dioxide have been captured and sequestered.

4. Advanced coal project credit (sec. 48A)

In general

An investment tax credit is available for power generation projects that use integrated gasification combined cycle ("IGCC") or other advanced coal-based electricity generation technologies.

First round of credit allocations

For the first round of credit allocations, the credit amount is 20 percent for investments in qualifying IGCC projects and 15 percent for investments in qualifying projects that use other advanced coal-based electricity generation technologies.

To qualify, an advanced coal project must be located in the United States and use an advanced coal-based generation technology to power a new electric generation unit or to retrofit or repower an existing unit. Generally, an electric generation unit using an advanced coal-based technology must be designed to achieve a 99-percent reduction in sulfur dioxide and a 90-percent reduction in mercury, as well as to limit emissions of nitrous oxide and particulate matter.[142]

The fuel input for a qualifying project, when completed, must use at least 75 percent coal. The project, consisting of one or more electric generation units at one site, must have a nameplate generating capacity of at least 400 megawatts, and the taxpayer must provide evidence that a majority of the output of the project is reasonably expected to be acquired or utilized.

Credits are available only for projects certified by the Secretary of Treasury, in consultation with the Secretary of Energy. Certifications are issued using a competitive bidding process. The Secretary of Treasury must establish a certification program no later than 180 days after August 8, 2005,[143] and each project application must be submitted during the three-year period beginning on the date such certification program is established. An applicant for certification has two years from the date the Secretary accepts the application to provide the Secretary with evidence that the requirements for certification have been met. Upon certification, the applicant has five years from the date of issuance of the certification to place the project in service.

The Secretary of Treasury may allocate $800 million of credits to IGCC projects and $500 million to projects using other advanced coal-based electricity generation technologies. Qualified projects must be economically feasible and use the appropriate clean coal technologies. With respect to IGCC projects, credit-eligible investments include only investments in property associated with the gasification of coal, including any coal handling and gas separation equipment. Thus, investments in equipment that could operate by drawing fuel directly from a natural gas pipeline do not qualify for the credit.

In determining which projects to certify, the Secretary must allocate power generation capacity in relatively equal amounts to projects that use bituminous coal, subbituminous coal, and lignite as primary feedstock. In addition, the Secretary must give high priority to projects which include greenhouse gas capture capability, increased by-product utilization, and other benefits.

Second round of credit allocations

Under the second round of credit allocations, the credit rate is increased to 30 percent for new IGCC and other advanced coal projects and the Secretary is permitted to allocate an additional $1.25 billion of credits to qualifying projects.[144]

Under the second round of credit allocations, qualifying projects must include equipment which separates and sequesters at least 65 percent of the project's total carbon dioxide emissions. This percentage increases to 70 percent if the credits are later reallocated by the Secretary. The Secretary is required to recapture the benefit of any allocated credit if a project fails to attain or maintain these carbon dioxide separation and sequestration requirements.

In selecting projects, the Secretary must give high priority to applicants who have a research partnership with an eligible educational institution. In addition, the Secretary must give the highest priority to projects with the greatest separation and sequestration percentage of total carbon dioxide emissions. The Secretary must also disclose which projects receive credit allocations, including the identity of the taxpayer and the amount of the credit awarded.

5. Gasification investment credit (sec. 48B)

In general

An investment credit is available for certain qualifying gasification projects. Only property which is part of a qualifying gasification project and necessary for the gasification technology of such project is eligible for the gasification credit.

First round of credit allocations

Under the first round of credit allocations, a 20-percent investment tax credit is available for investments in certain qualifying gasification projects. Qualified gasification projects convert coal, petroleum residue, biomass, or other materials recovered for their energy or feedstock value into a synthesis gas composed primarily of carbon monoxide and hydrogen for direct use or subsequent chemical or physical conversion. Qualified projects must be carried out by an eligible entity, defined as any person whose application for certification is principally intended for use in a domestic project which employs domestic gasification applications related to (1) chemicals, (2) fertilizers, (3) glass, (4) steel, (5) petroleum residues, (6) forest products, and (7) agriculture, including feedlots and dairy operations.

Credits are available only for projects certified by the Secretary of Treasury, in consultation with the Secretary of Energy. Certifications are issued using a competitive bidding process. The Secretary of Treasury must establish a certification program no later than 180 days after August 8, 2005,[145] and each project application must be submitted during the 3-year period beginning on the date such certification program is established. The Secretary of Treasury may not allocate more than $350 million in credits. In addition, the Secretary may certify a maximum of $650 million in qualified investment as eligible for credit with respect to any single project.

Second round of credit allocations

Under the second round of credit allocations, the gasification project credit rate is increased to 30 percent and the Secretary is authorized to allocate an additional $250 million of credits to qualified projects that separate and sequester at least 75 percent of total carbon dioxide emissions.[146] For this round, the definition of credit-eligible entities includes entities whose gasification projects are related to the production of transportation grade liquid fuels. The Secretary is required to recapture the benefit of any allocated credit if a project fails to attain or maintain these carbon dioxide separation and sequestration requirements.

In selecting projects, the Secretary must give high priority to applicants who have a research partnership with an eligible educational institution. In addition, the Secretary must give the highest priority to projects with the greatest separation and sequestration percentage of total carbon dioxide emissions. The Secretary must also disclose which projects receive credit allocations, including the identity of the taxpayer and the amount of the credit awarded.

6. Pollution Control Facilities (sec. 169)

In general, a taxpayer may elect to recover the cost of any certified pollution control facility over a period of 60 months.[147] A certified pollution control facility is defined as a new, identifiable treatment facility which (1) is used in connection with a plant in operation before January 1, 1976, to abate or control water or atmospheric pollution or contamination by removing, altering, disposing, storing, or preventing the creation or emission of pollutants, contaminants, wastes or heat; and (2) does not lead to a significant increase in output or capacity, a significant extension of useful life, a significant reduction in total operating costs for such plant or other property (or any unit thereof), or a significant alteration in the nature of a manufacturing production process or facility. Certification is required by appropriate State and Federal authorities that the facility complies with appropriate standards.

For a pollution control facility with a useful life greater than 15 years, only the portion of the basis attributable to the first 15 years is eligible to be amortized over a 60-month period.[148] In addition, a corporate taxpayer must reduce the amount of basis otherwise eligible for the 60- month recovery by 20 percent.[149] The amount of basis not eligible for 60-month amortization is depreciable under the regular tax rules for depreciation.

A certified air pollution control facility (but not a water pollution control facility) used in connection with an electric generation plant which is primarily coal fired will be eligible for 84- month amortization if the associated plant or other property was not in operation prior to January 1, 1976.[150] The 60-month amortization period remains in effect for any certified air pollution control facility used in connection with an electric generation plant which is primarily coal fired and which was in operation prior to January 1, 1976.

7. Special rules for nuclear decommissioning costs (sec. 468A)

In general

Special rules dealing with nuclear decommissioning reserve funds were enacted in the Deficit Reduction Act of 1984 ("1984 Act"), when tax issues regarding the time value of money were addressed generally. Under general tax accounting rules, a deduction for accrual basis taxpayers is deferred until there is economic performance for the item for which the deduction is claimed. However, the 1984 Act contains an exception under which a taxpayer responsible for nuclear power plant decommissioning may elect to deduct contributions made to a qualified nuclear decommissioning fund for future decommissioning costs. Taxpayers who do not elect this provision are subject to general tax accounting rules.

Qualified nuclear decommissioning fund

A qualified nuclear decommissioning fund (a "qualified fund") is a segregated fund established by a taxpayer that is used exclusively for the payment of decommissioning costs, taxes on fund income, management costs of the fund, and for making investments. The income of the fund is taxed at a reduced rate of 20 percent.

At the election of the taxpayer, a deduction is allowed for any taxable year in the amount of payments made by the taxpayer to a qualified fund. Amounts withdrawn by the taxpayer to pay for decommissioning costs are included in the taxpayer's income, but the taxpayer also is entitled to a deduction for decommissioning costs as economic performance for such costs occurs.

Accumulations in a qualified fund are limited to the amounts sufficient to cover the present value of a nuclear power plant's estimated decommissioning costs. In order to prevent accumulations of funds over the useful life of a nuclear power plant in excess of those required to pay decommissioning costs of such nuclear power plant and to ensure that contributions to a qualified fund are not deducted more rapidly than level funding (taking into account an appropriate discount rate), taxpayers must obtain a ruling from the IRS to establish the maximum annual contribution that may be made to a qualified fund (the "ruling amount"). In certain instances (e.g., change in estimates), a taxpayer is required to obtain a new ruling amount to reflect updated information.

A qualified fund may be transferred in connection with the sale, exchange or other transfer of the nuclear power plant to which it relates. If the transferee is a regulated public utility and meets certain other requirements, the transfer will be treated as a nontaxable transaction. No gain or loss will be recognized on the transfer of the qualified fund and the transferee will take the transferor's basis in the fund. The transferee is generally required to obtain a new ruling amount from the IRS.

G. Provisions Relating to General Energy-Related Projects and Research

1. Energy research credit (sec. 41)

General rule

A 20-percent credit is available for a taxpayer's expenditures on qualified research undertaken by an energy research consortium. Generally, an energy research consortium is an organization (not including a private foundation) that is either (1) a tax-exempt section 501 (c)(3) corporation organized and operated primarily to conduct energy research or (2) any other organization organized and operated primarily to conduct energy research in the public interest (within the meaning of section 501(c)(3)). In addition, an energy research consortium must have received payments or contributions for energy research from at least 5 unrelated persons per year and no single person can have paid or contributed a majority of the funds used for such research.

The energy research credit expires for amounts paid or incurred after December 31, 2009.[151]

Qualified research

Qualified research expenses eligible for the research tax credit consist of: (1) in-house expenses of the taxpayer for wages and supplies attributable to qualified research; (2) certain time-sharing costs for computer use in qualified research; and (3) 65 percent of amounts paid or incurred by the taxpayer to certain other persons for qualified research conducted on the taxpayer's behalf (so-called contract research expenses).[152] Notwithstanding the limitation for contract research expenses, qualified research expenses include 100 percent of amounts paid or incurred by the taxpayer to an eligible small business, university, or Federal laboratory for qualified energy research.

To be eligible for the credit, the research not only has to satisfy the requirements of present-law section 174 (described below) but also must be undertaken for the purpose of discovering information that is technological in nature, the application of which is intended to be useful in the development of a new or improved business component of the taxpayer, and substantially all of the activities of which constitute elements of a process of experimentation for functional aspects, performance, reliability, or quality of a business component. Research does not qualify for the credit if substantially all of the activities relate to style, taste, cosmetic, or seasonal design factors.[153] In addition, research does not qualify for the credit: (1) if conducted after the beginning of commercial production of the business component; (2) if related to the adaptation of an existing business component to a particular customer's requirements; (3) if related to the duplication of an existing business component from a physical examination of the component itself or certain other information; or (4) if related to certain efficiency surveys, management function or technique, market research, market testing, or market development, routine data collection or routine quality control.[154] Research does not qualify for the credit if it is conducted outside the United States, Puerto Rico, or any U.S. possession.

Relation to deduction

Under section 174, taxpayers may elect to deduct currently the amount of certain research or experimental expenditures paid or incurred in connection with a trade or business, notwithstanding the general rule that business expenses to develop or create an asset that has a useful life extending beyond the current year must be capitalized.[155] However, deductions allowed to a taxpayer under section 174 (or any other section) are reduced by an amount equal to 100 percent of the taxpayer's research tax credit determined for the taxable year.[156] Taxpayers may alternatively elect to claim a reduced research tax credit amount under section 41 in lieu of reducing deductions otherwise allowed.[157]

2. Credit for investment in advanced energy property (sec. 48C)

A 30-percent credit for investment in qualified property used in a qualified advanced energy manufacturing project. A qualified advanced energy project is a project that re-equips, expands, or establishes a manufacturing facility for the production: (1) property designed to be used to produce energy from the sun, wind, or geothermal deposits (within the meaning of section 61 3(e)(2)), or other renewable resources; (2) fuel cells, microturbines, or an energy storage system for use with electric or hybrid-electric motor vehicles; (3) electric grids to support the transmission of intermittent sources of renewable energy, including storage of such energy; (4) property designed to capture and sequester carbon dioxide; (5) property designed to refine or blend renewable fuels (but not fossil fuels) or to produce energy conservation technologies (including energy-conserving lighting technologies and smart grid technologies); (6) new qualified plug-in electric drive motor vehicles, qualified plug-in electric vehicles, or components which are designed specifically for use with such vehicles, including electric motors, generators, and power control units, or (7) other advanced energy property designed to reduce greenhouse gas emissions as may be determined by the Secretary. Qualified property does not include property designed to manufacture equipment for use in the refining or blending of any transportation fuel other than renewable fuels.

Qualified property must be depreciable (or amortizable) property used in a qualified advanced energy project. Only tangible personal property and other tangible property (not including a building or its structural components) are credit-eligible. The basis of qualified property must be reduced by the amount of credit received.

Credits are available only for projects certified by the Secretary of Treasury, in consultation with the Secretary of Energy. The Secretary of Treasury must establish a certification program no later than 180 days after February 17, 2009 (the date of enactment of the American Recovery and Reinvestment Act of 2009), and may allocate up to $2.3 billion in credits. No credit is allowed for any qualified investment that was allowed a credit under sections 48, 48A, or 48B.

In selecting projects, the Secretary may consider only those projects where there is a reasonable expectation of commercial viability. In addition, the Secretary must consider other selection criteria, including which projects (1) will provide the greatest domestic job creation; (2) will provide the greatest net impact in avoiding or reducing air pollutants or anthropogenic emissions of greenhouse gases; (3) have the greatest potential for technological innovation and commercial deployment; (4) have the lowest levelized cost of generated or stored energy, or of measured reduction in energy consumption or greenhouse gas emission; and (5) have the shortest project time from certification to completion.

Each project application must be submitted during the two-year period beginning on the date such certification program is established. An applicant for certification has one year from the date the Secretary accepts the application to provide the Secretary with evidence that the requirements for certification have been met. Upon certification, the applicant has three years from the date of issuance of the certification to place the project in service. Not later than four years after the date of enactment of the credit, the Secretary is required to review the credit allocations and redistribute any credits that were not used either because of a revoked certification or because of an insufficient quantity of credit applications.

H. Miscellaneous Energy Provisions

1. Tax-exempt bonds for power facilities (secs. 103, 141, and 142)

In general

Interest on bonds issued by State and local governments generally is excluded from gross income for Federal income tax purposes.[158] Because the interest income is excluded from gross income, investors generally are willing to accept a lower rate on on tax-exempt bonds than they might otherwise accept on a taxable investment. Thus, issuers of such bonds receive an implicit Federal subsidy equal to the difference between the tax-exempt interest rate paid and the taxable rate that otherwise would be paid. In this way, the income exclusion lowers the cost of capital for State and local government. Activities that can be financed with tax-exempt bonds issued by State and local governments include electric power facilities (i.e., generation, transmission, distribution, and retailing).

Bonds issued by State and local governments may be classified as either governmental bonds or private activity bonds. Governmental bonds are bonds the proceeds of which are primarily used to finance governmental functions or which are repaid with governmental funds. Private activity bonds are bonds in which the State or local government serves as a conduit providing financing to nongovernmental persons. For these purposes, the term "nongovernmental person" generally includes the Federal Government and all other individuals and entities other than States or local governments. The exclusion from income for interest on State and local bonds does not apply to private activity bonds, unless the bonds are issued for certain permitted purposes ("qualified private activity bonds") and other Code requirements are met.

Private activity bond tests

Present law provides two tests for determining whether a State or local bond is in substance a private activity bond and, as a result, not a governmental bond, the private business test and the private loan test.[159]

Private business tests

Private business use and private payments result in State and local bonds being private activity bonds if both parts of the two-part private business test are satisfied—

1. More than 10 percent of the bond proceeds is to be used (directly or indirectly) by a private business (the "private business use test"); and
2. More than 10 percent of the debt service on the bonds is secured by an interest in property to be used in a private business use or to be derived from payments in respect of such property (the "private payment test").[160]

Private business use generally includes any use by a business entity (including the Federal government), which occurs pursuant to terms not generally available to the general public. For example, if bond-financed property is leased to a private business (other than pursuant to certain short-term leases for which safe harbors are provided under Treasury regulations), bond proceeds used to finance the property are treated as used in a private business use, and rental payments are treated as securing the payment of the bonds. Similarly, in the case of public power entities, if output of an electric generating plant or transmission or distribution facilities is provided to a private business on terms not generally available to other customers of the entity, an allocable portion of bonds financing the facilities is treated as used in a private business use and as secured by the payments from the private business.

Private business use also can arise when a governmental entity contracts for the operation of a governmental facility by a private business under a management contract that does not satisfy Treasury regulatory safe harbors regarding the types of payments made to the private operator and the length of the contract.[161] These rules require public power entities to restrict the period of contracts with private businesses as well as the aggregate amount of electric service provided to private businesses on terms that are not generally available to customers of the entity, if interest on their bonds is to remain tax-exempt.

Private loan test

The second standard for determining whether a State or local bond is a private activity bond is whether an amount exceeding the lesser of (1) five percent of the bond proceeds or (2) $5 million is used (directly or indirectly) to finance loans to private persons. Private loans include both business and other (e.g., personal) uses and payments by private persons; however, in the case of business uses and payments, all private loans also constitute private business uses and payments subject to the private business test. Present law provides that the substance of a transaction governs in determining whether the transaction gives rise to a private loan. In general, any transaction which transfers tax ownership of property to a private person is treated as a loan. In the context of public power, longer-term contracts for the sale of output may violate the private loan test, because these contracts have the substantive characteristics of a loan.

Under the Energy Policy Act of 2005, qualified natural gas contracts are not treated as private loans for purposes of the private business tests. A contract is a qualified natural gas contract if the volume of natural gas secured for any year covered by the bond-financed contract does not exceed the sum of (1) the average annual natural gas purchased (other than for resale) by customers of a governmental utility within the service area of the utility ("retail natural gas consumption") during a testing period, and (2) the amount of natural gas that is needed to fuel transportation of the natural gas to the governmental utility. For purposes of this rule, the testing period is the 5-calendar-year period immediately preceding the calendar year in which the bonds are issued. A retail customer is one who does not purchase natural gas for resale.

Special legislative rules for tax-exempt financing of governmental "output" facilities

In addition to the general private business use and payment tests, the Code includes specific provisions governing the issuance of governmental tax-exempt bonds to finance "output" facilities. Output facilities include facilities for electric and gas generation, transmission, and distribution.

$15 million limit on private business use.–Present law imposes an additional restriction on private business use of State or local government bonds whose proceeds are to be used to finance output facilities.[162] A bond is treated as issued to finance an output facility (and subject to this restriction) if five percent or more of the proceeds is to be used with respect to any output facility. Under this restriction, the 10-percent private business use and private payment tests in substance are phased down for facilities that receive more than $15 million in tax-exempt bond financing. Significantly, unlike most tax-exempt bond restrictions, which are determined on a bond-issue by bond-issue basis, this restriction is measured by reference to all outstanding tax- exempt financing from which a facility benefits.

Bonds for acquisition of existing output property per se private activity.–In general, any bond with respect to which five percent or more ($5 million if less) of the proceeds is to be used, directly or indirectly, by a governmental entity to acquire existing output property is per se a private activity bond.[163] As such, interest on the bond is taxable, unless the use of the acquired facility satisfies the provisions applicable to tax-exempt private activity bonds for the local furnishing of electricity (described below). The two-county (or a city and a contiguous county) service area requirement that applies to facilities for the local furnishing of electricity does not apply in this circumstance.

There are two exceptions to the rule regarding the acquisition of existing output property. First, the rule does not apply to bonds for the acquisition of existing facilities that will provide service in a "qualified service area" of the issuer. A qualified service area is defined as an area throughout which the acquiring entity has provided electric service for at least the 10-year period preceding the date of the acquisition. Second, the rule does not apply to bonds issued to acquire existing output property to be used in a "qualified annexed area" of a public power entity. The term qualified annexed area includes only areas (1) that are contiguous to existing service areas, (2) that are annexed for general governmental purposes, and (3) the size of which does not exceed 10 percent of the public power entity's service area before the annexation occurs.

In addition, under the Energy Policy Act of 2005, qualified natural gas contracts (as defined above) are not treated as nongovernmental output property for purposes of the restrictions on the acquisition of such property.

Qualified private activity bonds

In general

As stated above, interest on private activity bonds is taxable unless the bonds meet the requirements for qualified private activity bonds. Qualified private activity bonds permit States or local governments to act as conduits providing tax-exempt financing for certain private activities. In most cases, the aggregate volume of these tax-exempt private activity bonds is restricted by annual aggregate volume limits imposed on bonds issued by issuers within each State. The Code further imposes several additional restrictions on tax-exempt private activity bonds that do not apply to bonds for governmental activities.

The definition of qualified private activity bonds includes an exempt facility bond, or qualified mortgage, veterans' mortgage, small issue, redevelopment, 501(c)(3), or student loan bond (sec. 141(e)). The definition of exempt facility bond includes bonds issued to finance certain transportation facilities (airports, ports, mass commuting, and high-speed intercity rail facilities); qualified residential rental projects; privately owned and/or operated utility facilities (sewage, water, solid waste disposal, and local district heating and cooling facilities, certain private electric and gas facilities, and hydroelectric dam enhancements); public/private educational facilities; qualified green building and sustainable design projects; and qualified highway or surface freight transfer facilities (sec. 142(a)).

Private activity bonds for the local furnishing of electricity

Qualified private activity bonds may be issued by States or local governments acting as conduits to finance generation, transmission, and distribution facilities for private businesses engaged in the local furnishing of electricity ("local furnishers"). A business is treated as engaged in local furnishing of electricity if the service territory in which the electricity is provided does not exceed (1) two contiguous counties, or (2) a city and a contiguous county. Historically, local furnishers eligible for this tax-exempt financing have included both investor- owned utilities ("IOUs") and independent power ventures. These bonds may be issued for the benefit of only those persons that were engaged in local furnishing of electricity in the service territory in which the new facilities will be used as of January 1, 1997, or in qualified expansions of those service territories. A "qualified expansion" is limited to service territory that is a part of a county in which the local furnisher was providing electric service on that date. For example, if a local furnisher was providing electric service to one county and a portion of a contiguous county on January 1, 1997, bonds may be issued for the continued provision of service both within that area and also for service to be provided in the remaining portion of the contiguous county in the future. In addition to persons actually engaged in local furnishing activities on January 1, 1997, the Code allows certain successors in interest to persons that qualified as local furnishers on that date to "step into the shoes" of the predecessor local furnishers provided that the service territories served otherwise satisfy the requirements for local furnishing.

Notwithstanding the general limits on service territories of local furnishers, the Code includes special rules allowing these electric service providers to transmit ("wheel") electricity through their systems, if ordered by the Federal Energy Regulatory Commission to do so under sections 211 or 213 of the Federal Power Act, provided that the size of the transmission lines or other facilities used in these wheeling activities does not exceed the capacity required to serve their otherwise qualified two contiguous county service area or city and contiguous county service area.

Hydro-electric generating facilities

Present law also permits the issuance of qualified private activity bonds for "environmental enhancements of hydro-electric generating facilities." Eligible facilities include those that protect or promote fisheries or other wildlife resources and those for recreational purposes or other improvements required by the terms of a Federal license for the operation of a hydro-electric generating facilities. This provision was enacted to permit tax-exempt financing of certain renovations to the dams and accompanying hydroelectric electric generating facilities along the Columbia River that are a part of the Bonneville Power Administration system. Bonds issued for these purposes are not subject to the State volume limitations applicable to most qualified private activity bonds.

2. Reduced rate of tax for diesel-water fuel emulsion (secs. 4081(a)(2)(D), 4081(c), and 6427(m))

Diesel fuel is taxed at 24.3 cents per gallon. A special tax rate of 19.7 cents per gallon is provided for diesel fuel blended with water into a diesel-water fuel emulsion to reflect the reduced Btu content per gallon resulting from the water. Emulsion fuels eligible for the special rate must consist of at least 14 percent water. The emulsion additive must be registered by a United States manufacturer with the Environmental Protection Agency pursuant to section 211 of the Clean Air Act (as in effect on March 31, 2003). A refund of the difference between the regular rate for diesel fuel (24.3 cents per gallon) and the incentive rate for diesel-water emulsions (19.7 cents per gallon) is available to the extent tax-paid diesel is used to produce a qualifying emulsion diesel fuel. Anyone who separates the diesel fuel from the diesel-water fuel emulsion on which a reduced rate of tax was imposed is treated as a refiner of the fuel and is liable for the difference between the amount of tax on the latest removal of the separated fuel and the amount of tax that was imposed upon the pre-mixture removal. The special tax rate does not apply if the person removing, selling or using such fuel is not registered with the IRS.

3. Exceptions for publicly traded partnerships treated as corporations (secs. 7704 and 851)

Treatment as a corporation

Present law provides that a publicly traded partnership means a partnership, interests in which are traded on an established securities market, or are readily tradable on a secondary market (or the substantial equivalent thereof). In general, a publicly traded partnership is treated as a corporation, but an exception to corporate treatment is provided if 90 percent or more of its gross income is interest, dividends, real property rents, or certain other types of qualifying income.[164]

Under this exception, income and gains from certain activities with respect to minerals or natural resources are treated as qualifying income. Specifically, among other types of income, qualifying income includes income and gains derived from the exploration, development, mining or production, processing, refining, transportation (including pipelines transporting gas, oil, or products thereof), or the marketing of any mineral or natural resource (including fertilizer, geothermal energy, and timber). For this purpose, oil, gas, or products thereof means gasoline, kerosene, number 2 fuel oil, refined lubricating oils, diesel fuel, methane, butane, propane, and similar products which are recovered from petroleum refineries or field facilities. Oil, gas, or products thereof are not intended to encompass oil or gas products that are produced by additional processing beyond that of petroleum refineries or field facilities, such as plastics or similar petroleum derivatives.

Qualifying income for regulated investment companies

A regulated investment company (RIC) generally is treated as a conduit for Federal income tax purposes. To qualify for conduit treatment, at least 90 percent of the RIC's gross income must fall within specified categories (dividends, interest, and other specified types). Income derived from a partnership generally is treated as meeting this requirement only to the extent such income is attributable to items of income of the partnership that would meet the requirement if realized by the RIC in the same manner as realized by the partnership (the "look- through" rule for partnership income).[165]

In the case of a publicly traded partnership, present law provides that the 90-percent test with respect to income of a RIC includes net income derived from an interest in a publicly traded partnership. The look-through rule for partnership income of a RIC applies only to income from a partnership other than a publicly traded partnership.

III. DATA AND ANALYSIS

Energy Tax Expenditures and Sources of U.S. Energy Supply

Table 9, below, shows the staff of the Joint Committee on Taxation's most recent estimates of the energy-specific tax expenditures. The largest tax expenditure items generally occur in the oil and gas industries, owing at least in part to the relative size of such industries as compared to the renewable energy sector.

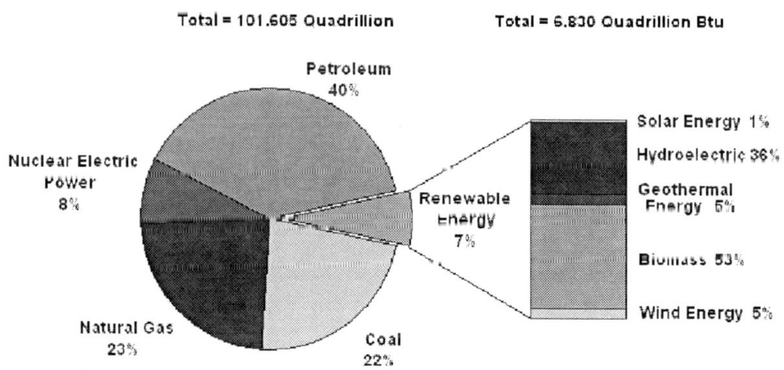

Source: Energy Information Administration, Renewable Energy Consumption and Electricity Preliminary 2007 Statistics, May 2008.

Chart 1. The Role of Renewable Energy Consumption in the Nation's Energy Supply, 2007

In recent years there has been increased interest in providing tax subsidies for conservation of energy and for development of renewable sources of energy. However, the United States continues to rely primarily on fossil fuel sources for energy. In 2007, 85 percent of U.S. energy supply came from fossil fuels, eight percent from nuclear electric power, and only seven percent from renewable sources of energy. Chart 1, below, shows the distribution of the U.S. energy supply by source of energy – petroleum, natural gas, coal, nuclear, and renewable.

General Economic Rationale for Certain Tax Expenditure Intervention in Energy Markets

A common economic rationale for government intervention in certain markets (including many aspects of energy markets) is that often there exist "externalities" in the consumption or production of certain goods that lead to "market failures," wherein either too little or too much of certain economic activity occurs relative to what is the socially optimal level of activity. An externality exists when, in the consumption or production of a good, there is a difference between the cost or benefit to an individual from consumption or production and the cost or benefit to society as a whole. When the society-wide, or "social" costs of consumption exceed the private costs of consumption, a negative externality exists. When the social benefits from consumption or production exceed private benefits, a positive externality exists. When negative externalities exist, there will be over-consumption of the good causing the negative externality relative to what would be socially optimal. When positive externalities exist, there will be under consumption of the good producing the positive externality. The reason for the over consumption or under consumption is that private actors will in general not take into account the effect of their consumption on others, but only weigh their personal costs and benefits in their decisions. Thus, they will consume goods up to the point where their marginal benefit of more consumption is equal to the marginal cost (generally, the price) that they face. But from a social perspective, consumption should occur up to the point where the marginal social cost (generally, the price to the consumer plus any external costs imposed on others) is equal to the marginal social benefit (the benefit received by the consumer, plus any social benefit from the individual consumption). Absent any government intervention, only when there are no externalities will the private actions lead to the socially optimal level of consumption or production, because only in this case will private costs and benefits be equal to social costs and benefits.

As a general matter of economic efficiency, tax preferences designed to influence investment choices should be used only when it is acknowledged that market-based pricing signals have led to a lower level of investment in a good than would be socially optimal. In general, this can occur in a market-based economy when private investors do not capture the full value of an investment–that is, when there are positive externalities to the investment that accrue to third parties who did not bear any of the costs of the investments. For example, if an individual or corporation can borrow funds at 10 percent and make an investment that will return 15 percent, they will generally make that investment. However, if the return were 15 percent, but only 8 percent of that return went to the investor, and seven percent to society at large, the investment will generally not take place, even though the social return (the sum of the return to the investor and other parties) would indicate that the investment should be made. In such a situation, it may be desirable to subsidize the return to the investor through tax credits or other mechanisms in order that the investor's return is sufficient to cause the socially desirable investment to be made. In this example, a credit that raises the return to the investor to at least 10 percent would be necessary. Even if the cost of the credit were paid through general tax increases for others, they would presumably be better off since they enjoy a seven percent return from the investment, and the credit would only need to raise the return to the investor by two percent for him or her to break even. Thus, even if the rest of society bears the full cost of the credit, they would enjoy a five-percent net return to the investment (seven percent less two percent).

Table 9. Energy-Related Tax Expenditure Estimates, Fiscal Years 2008 - 2012 [Billions of Dollars]

Function	Corporations					Individuals					Total
	2008	2009	2010	2011	2012	2008	2009	2010	2011	2012	2008-12
Energy related credits:											
Credit for enhanced oil recovery costs	[1]	[1]	[1]	[1]	[1]	[1]	[1]	[1]	[1]	[1]	0.2
Credit for producing fuels from a non-conventional source	0.1	0.1	[1]	[1]	[1]	[1]	[1]	[1]	[1]	[1]	0.6
Credits for alcohol fuels [2]	0.1	0.1	0.1	[1]	---	---	---	---	---	---	0.3
Energy credit (section 48):											
Solar	[1]	[1]	[1]	[1]	[1]	[1]	[1]	[1]	[1]	[1]	[1]
Geothermal	[1]	[1]	[1]	[1]	[1]	[1]	[1]	[1]	[1]	[1]	[1]
Fuel cells	[1]	[1]	[1]	[1]	[1]	[1]	[1]	[1]	[1]	[1]	[1]
Microturbines	---	---	---	---	---	[1]	[1]	[1]	[1]	[1]	[1]
Credits for electricity production from renewable resources (section 45):											
Wind	0.6	0.8	0.9	0.9	0.9	[1]	[1]	[1]	[1]	[1]	4.1
Closed-loop biomass	[1]	[1]	[1]	[1]	[1]	---	---	---	---	---	[1]
Geothermal	[1]	[1]	[1]	[1]	[1]	---	---	---	---	---	0.2
Qualified hydropower	[1]	[1]	[1]	[1]	[1]	---	---	---	---	---	0.1
Solar (limited to facilities placed in service before 1/1/06)	[1]	[1]	[1]	[1]	[1]	---	---	---	---	---	0.1
Small irrigation power	[1]	[1]	[1]	[1]	[1]	---	---	---	---	---	0.1
Municipal solid waste	[1]	[1]	[1]	[1]	[1]	---	---	---	---	---	0.2
Open-loop biomass	0.3	0.4	0.3	0.2	0.2	[1]	[1]	[1]	[1]	[1]	1.5
Credits for investments in clean coal facilities	0.2	0.2	0.2	0.1	0.1	---	---	---	---	---	0.8
Coal production credits:											
Refined coal	[1]	[1]	[1]	[1]	[1]	---	---	---	---	---	0.1
Indian coal	[1]	[1]	[1]	[1]	[1]	---	---	---	---	---	0.1
Credit for the production of energy-efficient appliances	0.1	0.1	0.1	[1]	---	---	---	---	---	---	0.3

Table 9. (Continued)

Function	Corporations					Individuals					Total
	2008	2009	2010	2011	2012	2008	2009	2010	2011	2012	2008-12
Credits for alternative technology vehicles:											
Hybrid vehicles	[1]	[1]	[1]	[1]	[1]	0.2	0.2	0.2	0.1	[1]	0.9
Other alternative fuel vehicles	[1]	[1]	[1]	[1]	[1]	[1]	[1]	[1]	[1]	[1]	0.1
Credit for clean-fuel vehicle refueling property	[1]	[1]	[1]	[1]	[1]	[1]	[1]	[1]	[1]	[1]	0.1
Residential energy efficient property credit	---	---	---	---	---	[1]	[1]	---	---	---	0.1
New energy efficient home credit	[1]	[1]	[1]	[1]	[1]	---	---	---	---	---	0.1
Credit for energy efficiency improvements to existing homes	---	---	---	---	---	0.8	0.3	1.0	---	---	2.1
Credits for alternative technology vehicles	0.1	[1]	[1]	[1]	[1]	0.2	0.2	0.2	[1]	[1]	1.0
Credit for holders of clean renewable energy bonds	[1]	0.1	0.1	0.1	0.1	[1]	[1]	[1]	[1]	[1]	0.5
Credit for holder of qualified energy conservation bonds	[1]	[1]	[1]	[1]	[1]	[1]	[1]	[1]	[1]	[1]	0.1
Energy-related exclusions from income:											
Exclusion of interest on State and local government qualified private activity bonds for energy production facilities	[1]	[1]	[1]	[1]	[1]	0.1	0.1	0.1	0.1	0.1	0.6
Exclusion of energy conservation subsidies provided by public utilities	---	---	---	---	---	[1]	[1]	[1]	[1]	[1]	0.1
Energy-related deductions:											
Deduction for expenditures on energy-efficient commercial building property	[1]	0.1	0.1	0.1	0.1	[1]	[1]	[1]	[1]	[1]	0.2
Eight-year inclusion from sale of electric transmission assets to independent utilities	0.3	0.2	[1]	-0.1	-0.1	[1]	[1]	[1]	[1]	[1]	0.3
Expensing of exploration and development costs:											
Oil and gas	2.1	3.0	1.7	0.4	0.4	[1]	[1]	[1]	[1]	[1]	7.2
Other fuels	[1]	[1]	[1]	[1]	[1]	[1]	[1]	[1]	[1]	[1]	0.3
Excess of percentage over cost depletion:											
Oil and gas	1.3	1.4	1.4	1.4	1.4	[1]	[1]	[1]	[1]	[1]	7.1
Other fuels	0.1	0.1	0.1	0.1	0.1	[1]	[1]	[1]	[1]	[1]	0.8
Amortization of geological and geophysical expenditures associated with oil and gas exploration	0.1	0.2	0.2	0.2	0.2	[1]	0.1	0.1	0.1	0.1	1.3
Amortization of air pollution control facilities	0.1	0.1	0.1	0.2	0.2	[1]	[1]	[1]	[1]	[1]	0.6

Table 9. (Continued)

Function	Corporations					Individuals					Total
	2008	2009	2010	2011	2012	2008	2009	2010	2011	2012	2008-12
Depreciation recovery periods for energy specific items:											
Five-year MACRS for certain energy property (solar, wind, etc.)..	0.2	0.1	0.1	0.1	0.1	[1]	[1]	[1]	[1]	[1]	0.5
10-year MACRS for smart electric distribution property	[1]	[1]	[1]	[1]	[1]	0.1	[1]	[1]	[1]	[1]	0.1
15-year MACRS for certain electric transmission property	[1]	0.1	0.1	0.2	0.2	[1]	[1]	[1]	[1]	[1]	0.6
15-year MACRS for natural gas distribution line	0.1	0.1	0.1	0.1	0.1	[1]	[1]	[1]	[1]	[1]	0.6
Election to expense 50 percent of qualified property used to refine liquid fuels	0.4	1.1	0.9	0.8	0.6	[1]	[1]	[1]	[1]	[1]	3.8
Exceptions for publicly traded partnership with qualified income derived from certain energy-related activities	---	---	---	---	---	0.4	0.4	0.5	0.6	0.6	2.6

Joint Committee on Taxation.

Estimates based on the January 2008 Congressional Budget Office baseline for tax expenditures under present law as of October 31, 2008.

NOTE: Details may not add to totals due to rounding.

[1] Positive tax expenditure of less than $50 million.

[2] In addition, the credit from excise tax for alcohol fuels results in a reduction in excise tax receipts, net of income tax effect, of $13.6 billion over the fiscal years 2008 through 2012.

Pollution is an example of a negative externality, because the costs of pollution are borne by society as a whole rather than solely by the polluters themselves. In the case of pollution, there are various ways the government could intervene in markets to limit pollution to more economically efficient levels. One approach is to control pollution directly through regulation of polluters, such as by requiring coal burning electric utilities to install scrubbers to limit their emissions of various pollutants. Other more market oriented approaches to achieving socially optimal levels of pollution control are also possible. One such approach is to set a tax on the polluting activity that is equal to the social cost of the pollution. Thus, if burning a gallon of gasoline results in pollution that represents a cost to society as a whole of $1, it would be economically efficient to tax gasoline at $1 per gallon. By so doing, the externality is said to be internalized, because now the private polluter faces a private cost equal to the full social cost, and the socially optimal amount of consumption will take place. An alternative market-based approach to control pollution is to employ a system of payments, such as perhaps tax credits, to essentially pay polluters to reduce pollution. If the payments can be set in such a way as to yield the right amount of reduction (that is, without paying for reduction more than the reduction is valued, or failing to pay for a reduction where the payment would be less than the value of the pollution reduction), the socially desirable level of pollution will result. The basic difference between these two approaches is a question of who pays for the pollution reduction. The tax approach suggests that the right to clean air is paramount to the right to pollute, as polluters and those who buy goods and services from polluters would bear the social costs of their pollution. The alternative approach suggests that the pollution reduction costs should be borne by those who receive the benefit of the reduction.

In the case of a positive externality, the appropriate economic policy would be to impose a negative tax (i.e., a tax preference) on the consumption or production that produces the positive externality. By the same logic as above, the externality becomes internalized, and the private benefits from consumption become equal to the social benefits, leading to the socially optimal level of consumption or production. An example where such a positive externality is thought to exist is in basic scientific research, as the social payoffs to such research are not fully captured by private parties that undertake, and incur the cost of, such research. As a result, a socially suboptimal level of such research is undertaken. The provision of a subsidy for such research can correct this market inefficiency and lead to socially optimal levels of research.

Some have also argued that decreasing the dependence of the U.S. on foreign source energy is desirable for geopolitical and national defense reasons, and that this provides a rationale for subsidizing domestic fossil fuel production as well as subsidizing conservation and renewable energy production. However, in recent years there has been increasing focus in the tax code on energy conservation and renewable energy production incentives. The remainder of the discussion herein will focus on some considerations in the design of these incentives.

Issues in the Design and Efficacy of Tax Expenditures for Energy Conservation and Renewable Energy Production

Economists generally agree that the most efficient means of addressing pollution would be through a direct tax on the pollution-causing activities, rather than through the indirect approach of targeted tax credits for certain technologies. By the direct tax approach, the establishment of the economically efficient prices on pollutants, through taxes, would result in the socially optimal level of pollution. To achieve this result, the tax would be set equal to the cost to society of the incremental pollution. The imposition of a direct tax on the pollution- causing activity would indirectly lead to the adoption of the types of technologies favored in the tax code, but only if these technologies were in fact the most socially efficient technologies. A tax on the pollution causing activity is technologically neutral–a tax does not favor any particular technology that individuals might choose to utilize, or favor any particular behavioral modification that individuals may choose to make, in their pollution reducing responses to the tax. Rather, individuals would choose the most cost effective and economically efficient means of altering their behavior in response to the tax. For example, the optimal behavioral responses to a broad based tax on fossil fuels would lead to installation of greater amounts of home insulation, but would also lead to individuals turning down the thermostat or switching off unnecessary lighting. It would be difficult or impractical to design tax subsidies to directly incentivize the turning down of thermostats, the switching off of lights, or other similar forms of optimizing behavior.

Nonetheless, many provisions of current law provide targeted tax credits for investment in, or expenditures on, certain assets that reduce, directly or indirectly, the consumption of conventional fuels and the attendant pollutants and emissions of gases related to atmospheric warming. The design of these tax benefits is important to how close they will come, individually and collectively, to achieving their intended objectives in a cost effective and efficient manner. Ideally, their design would be coordinated to try to mimic the more economically efficient outcome that a broad based tax would provide.

The most important consideration in the efficient design of targeted subsidies is to determine what activities to subsidize and how much to subsidize them (i.e., what a credit rate should be, for example). In setting the policy parameters, the government is implicitly setting the price it will pay for the energy production or conservation that is produced or conserved in the manner specified by the tax provision. To be technologically neutral and economically efficient, the government should seek to set its policy parameters so that the implicit price it pays for the same good, say fossil fuel displacement (typically measured in millions of British thermal units,[166] or "MMBtu's"), is the same under each tax provision that has the same purpose. If it sets its policies in this manner, then only the most cost effective production of such fossil fuel MMBtu displacement will be subsidized.

While the government's policy parameters indicate the price it is willing to pay for fossil fuel MMBtu displacement at the margin, in practice it is difficult to know how much overall incremental fossil fuel displacement (pollution reduction) the government is buying in the aggregate with a given conservation or renewables production credit. The reason is that the government subsidy typically applies to "inframarginal" activity, or activity that would have occurred even in the absence of the credit, so for such activity the government incurs an expense in subsidizing it in order to induce others at the margin to engage in the tax-favored activity. For example, a credit exists for the purchase of certain energy efficient furnaces. If we assume the credit were $500 and assume that the typical energy consumption from the efficient furnace as compared to an average furnace resulted in 1,000 MMBtu less fossil fuel consumption over its lifetime, then government has set the price of 50 cents for each MMBtu of displaced fossil fuel consumption to encourage the adoption of the more efficient furnace. However, many investments in the energy efficient furnaces might have taken place even in the absence of the credit, and thus the government pays, via the credit, for fossil fuel displacement that would have occurred anyway. If two million furnaces are sold (leading to a billion dollars in credits being claimed), but only 200,000 of these were sold as a direct result of the credit, then only one tenth (200,000 divided by 2 million) of the fossil fuel displacement from the energy efficient furnaces can be said to have occurred because of the credit. Thus, in this hypothetical example, the true budget cost of the aggregate incremental displaced fossil fuel consumption is 10 times the implicit government price at the margin, or $5, for each MMBtu of displaced fossil fuel consumption.[167] Additionally, individuals who have purchased a more efficient furnace might choose to heat their home to a greater degree since it will cost less to do so, thus negating some of the initial fossil fuel displacement from the purchase of the more efficient furnace, and further inflating the cost to the government of a given amount of fossil fuel displacement.[168]

While the government can in theory establish an efficient set of subsidies for the activities it chooses to subsidize, in practice it cannot administratively identify and set up programs to subsidize every conceivable energy saving practice. Additionally, it is not possible to identify meritorious technologies not yet invented, and the government must continue to expand the class of credit-eligible activities if it wishes to minimize the economic distortions that come from favoring certain technologies through tax subsidies over other technologies that prove equally capable of achieving reductions in fossil fuel consumption. Furthermore, the investment in research to develop such new technologies might be constrained by the existence of tax subsidies for current technologies, as investors in such research run the political risk that their newly discovered technologies will not be granted any tax subsidies in order to compete favorably with subsidized technologies.

Table 10 compares the implicit price that the government is willing to pay per MMBtu of fossil fuel displacement of selected tax subsidies. The differing amounts paid for a given amount of MMBtu show that the government's policies are not coordinated to pay the same price for MMBtu displacement regardless of source, and thus at the margin the government pays more to purchase MMBtu displacement from certain activities over other activities, which is not economically efficient.[169] Column 1 in Table 10 lists the statutory credit amount in cents per kilowatt-hour and dollars or cents per gallon. Column 2 normalizes the statutory credit amounts to express them in terms of dollars per unit of heat energy (in millions of Btu's) embedded in the fuel or in the kilowatt-hour of electricity. While a renewable fuel, such as ethanol, directly displaces a fossil fuel on a Btu per Btu basis, the fossil fuel heat energy that a kilowatt-hour of renewable electricity displaces (because a fossil fuel plant will not have to generate that kilowatt- hour) will depend on the thermal efficiency with which a fossil-fueled electricity generation station converts the heat energy of the fossil fuel to the heat energy of a kilowatt-hour of electricity. This measure of the generating station's thermal efficiency is known as the "heat rate." According to the Department of Energy, the average annual heat rate factor for fossil- fueled power plants in the United States in 2007 was 9,884 Btus per kilowatt-hour. Thus, though a kilowatt-hour of electricity has heat energy of 3,412 Btus, as noted at the bottom of Table 10, it requires on average 9,884 Btus of fossil fuel to produce that kilowatt-hour at a domestic fossilfuel-burning power plant. Thus, a kilowatt-hour of renewable electricity displaces on average 9,884 Btus of fossil fuel feedstock. Column 3 in Table 10 shows the credit amount per million Btus of displaced fossil fuel consumption. If the objective of the federal government's renewable energy policy is defined as displacement of fossil fuel energy, then column 3 shows the varying amounts that the government pays to accomplish that objective.

As noted above, it cannot be known from this information alone what the total budget cost of the aggregate *incremental* renewable production that occurs as a result of the credits, due to renewable production that would have occurred in the absence of the credits. If, as an example, half of the wind energy production would have occurred in any event, then the total federal revenue cost of achieving the incremental wind energy produced is twice that stated in the table, if one assumes that all wind energy produced receives the credit.[170]

One can also compute the implicit price that the government is willing to pay per MMBtu for the various provisions designed to encourage taxpayers to conserve energy. As an example, Table 11, below, computes the implicit price the government is willing to pay to conserve motor fuel (normalized to MMBtu to facilitate comparison to Table 10, above) in the case of three hybrid motor vehicles available for purchase in 2006 for which taxpayers could claim a tax credit. For each vehicle listed in Table 11 the EPA estimated fuel economy for the vehicle was compared to a comparable non-hybrid vehicle and the lifetime fuel saving was calculated (third column).[171] The last column reports the estimated implicit price the government paid to conserve one million Btu by dividing the maximum eligible tax credit a taxpayer could claim on the hybrid motor vehicle by the amount of energy saved over the life of the vehicle.

As was the case for the energy production tax credits reviewed in Table 10, Table 11 shows that the credits for the purchase of hybrid motor vehicles were not coordinated to pay the same price of MMBtu displacement regardless of source, and thus at the margin the government would pay more to purchase MMBtu displacement from certain vehicles over other vehicles, which creates an inefficiency to the government's tax expenditure.

Table 10. Comparison of Selected Energy Production Tax Credits

	(1) Statutory credit amount	(2) Credit amount in dollars per MMBtus of heat energy	(3) Credit amount in dollars per MMBtus of displaced heat energy of fossil fuel feedstock
Wind power	2.1 cents per kilowatt-hour	$6.15	$2.12
Geothermal power	2.1 cents per kilowatt-hour	$6.15	$2.12
Open loop biomass	1 cent per kilowatt-hour	$2.93	$1.01
Advanced nuclear power	1.8 cents per kilowatt-hour	$5.28	$1.82
Ethanol	45 cents per gallon	$5.92	$5.92
Biodiesel	$1 per gallon	$8.45	$8.45

Notes:
1 kilowatt-hour = 3,412 Btus
1 gallon of ethanol = 76,000 Btus (low heating value)
1 gallon of biodiesel = 118,296 Btus (low heating value)
Displaced fossil fuel feedstock calculation assumes a fossil fuel heat rate thermal conversion factor for wind, geothermal, biomass, and nuclear power of 9,884 Btus per kilowatt-hour.
Btus per kw-hour and thermal heat rate conversion factor taken from Energy Information Agency, Monthly Energy Review, Table A6 (April 2009)
Btu content of ethanol and biodiesel taken from Energy Information Agency, Annual Energy Outlook, Table 12 (2007)

Table 11. Estimated Tax Credit Per MMBtu Conserved by Purchase of Selected Hybrid Motor Vehicles

Vehicle	Maximum Eligible Tax Credit	Estimated Lifetime Fuel Saving	Dollars of Tax Credit Per MMBtu
2007 Toyota Camry 2.4 L four-door sedan	$2,600	39 barrels of oil	$11.49
2006 Honda Accord V6 AT four-door sedan	$1,300	34 barrels of oil	$6.59
2006 GMC Sierra 1500 SL pickup four wheel drive pickup truck	$650	20 barrels of oil	$5.60

Similar calculations can be made for other tax preferences that are intended to encourage conservation or displace existing energy sources with more environmentally benign energy sources. However, many such calculations are sensitive to the geographic location of the taxpayer and the qualified energy property. The payoff in reduced energy consumption from additional insulation of a personal residence depends upon the climate in which the taxpayer resides and the amount of insulation initially in the residence. The tax credit available to taxpayers for additional insulation depends only upon the quantity of insulation and the price paid for the insulation, and the price of insulation does not vary widely across the nation. Therefore, the implicit price that the government is willing to pay per MMBtu conserved will vary with such factors as the location of the taxpayer and pre-existing levels of insulation. As a further example, consider the hypothetical installation of a 10-kilowatthour rated photovoltaic power system. The University of California Energy Institute estimates the installed cost of such a system at approximately $80,000.[172] If, over the assumed 25-year life of such a system, it could garner eight hours of daylight for 365 days per year, it would produce 730,000 kilowatt- hours of electricity offsetting as much electricity produced from other sources. The present-law 30-percent tax credit for the installation of such a system would imply that the government was willing to pay $9.64[173] per MMBtu of displaced electricity. However, if the same system were only to average five hours of sunlight per day, it would produce 456,250 kilowatt-hours of electricity. As the installed cost of the system does not vary, in this case the present-law 30- percent tax credit for the installation of the system would imply that the government was willing to pay $15 .42 [174] per MMBtu of displaced electricity.

Alternative minimum tax, nonrefundability, and other constraints on tax expenditures

Another design issue that affects the efficacy of many tax credits is their restricted availability. Many tax credits have stipulated dollar limitations, are nonrefundable, or cannot be used to offset tax liability determined under the alternative minimum tax ("AMT"). If a credit designed to overcome an externality is capped, then after the cap is reached the marginal cost of further investment becomes equal to the market price again, which is presumed to be inefficient because of the externality. The impact of these limitations is to make the credit less valuable to those without sufficient tax liability to claim the full credit, for those subject to the AMT, or those who have reached any cap on the credit. Given the arguments outlined above as to the rationale for targeted tax credits, it is not economically efficient to limit their availability based on the tax status of a possible user of the credit. It can be argued that, if such social benefits exist and are best achieved through the tax system, the credit should be both refundable and available to AMT taxpayers. In some cases making the credits refundable may introduce compliance problems that would exceed the benefits from encouraging the targeted activities for the populations lacking sufficient tax liability to make use of the credit. With respect to the AMT, the rationale for the limitation is to protect the objective of the AMT, which is to insure that all taxpayers pay a minimum (determined by the AMT) amount of tax. Two differing policy goals thus come in conflict in this instance. Similarly, caps on the aggregate amount of a credit that a taxpayer may claim are presumably designed to limit the credit's use out of some sense of fairness, but again, this conflicts with the goal of pollution reduction.

End Notes

[1] This document may be cited as follows: Joint Committee on Taxation, *Tax Expenditures for Energy Production and Conservation* (JCX-25-09R) April 21, 2009. This document can also be found on our website at www.jct.gov.

[2] Sec. 30C.

[3] See secs. 40, 40A, 6426, and 6427(e). In addition, the payments authorized under section 6427(e) of the Code for alcohol fuel mixtures, biodiesel fuel mixtures, renewable diesel fuel mixtures, alternative fuel mixtures and alternative fuel, may be taken as a refundable income tax credit pursuant to section 34 and section 6401(b)(1) of the Code. The section 34 income tax credit is not available for any amount for which a timely claim has been filed under section 6427.

[4] For the section 40 income tax credit, there is a smaller credit amount for alcohol with a proof of at least 150 but less than 190.

[5] Before January 1, 2009, the Code permitted the volume of alcohol eligible for the credit to include up to five percent denaturants. Gasoline is often used as denaturant. Section 15332 of the Food, Conservation and Energy Act of 2008 (Pub. L. No 110-234) reduced allowable denaturants from five percent to two percent. Under Alcohol, Tobacco Tax and Trade Bureau regulations (27 C.F.R. sec. 19.1005) for alcohol to be eligible for withdrawal as fuel alcohol, denaturants included in the volume of the fuel alcohol must exceed 1.96 percent of the volume of the fuel alcohol. Because of comments from the public about the ability to accurately measure the volume of denaturants in alcohol, Notice 2009-6, 2009-3 I.R.B. (2009), provides a safe harbor and transitional rule to implement the two-percent limitation. The IRS will treat denaturants as not clearly exceeding the two-percent limitation unless there is clear evidence establishing that the denaturants exceed 2.5 percent of volume (including denaturants). The safe harbor is available only if no more than two percent of the alcohol's volume is disregarded under the rule providing that proof is determined without regard to added denaturants.

[6] For this purpose, "United States" includes any possession of the United States.

[7] See secs. 40A(d)(1), 40A(f)(3), 6426(h).

[8] Sec. 40A.

[9] Sec. 6426(c).

[10] Sec. 6426(c)(4).

[11] Sec. 6427(e).

[12] Secs. 40A(f), 6426(c), and 6427(e).

[13] "Gasoline gallon equivalent" means, with respect to any nonliquid alternative fuel (for example, compressed natural gas), the amount of such fuel having a Btu (British thermal unit) content of 124,800 (higher heating value).

[14] See Internal Revenue Service, Notice 2006-92, *Alternative Fuel, Alternative Fuel Mixtures; Blood Collector Organizations*, 2006-43 I.R.B. 774 (October 23, 2006). Recent alternative fuel mixture claims for the burning of black liquor (a byproduct of the manufacturing of pulp and paper) and diesel fuel in recovery boilers have highlighted the fact that the fuel mixture credits are not limited to transportation uses, and that there is no upper limit on the dollar amount that may be claimed. See, International Paper, *News Release: International Paper Provides Update on Alternative Fuel Credits* (March 24, 2009); Steve Mufson, Washington Post, *Papermakers Dig Deep in Highway Bill to Hit Gold* (March 28, 2009 at p. D.1); Rebecca Penty, New Brunswick Business Journal, *Canada at a Disadvantage: Forestry Industry Contends Credits are Subsidies that Allow U.S. Firms to Better Compete* (April 3, 2009 at p. B.1); and Jad Mouawad and Clifford Krauss, The New York Times, *Lawmakers May Limit Paper Mills' Windfall* (April 18, 2009).

[15] Sec. 45. In addition to the renewable electricity production credit, section 45 also provides income tax credits for the production of Indian coal and refined coal at qualified facilities. A grant program is also available for 2009 and 2010 allowing taxpayers to apply for a thirty-percent grant in lieu of the renewable electricity production credit.

[16] Sec. 38(b)(8).

[17] Sec. 38(c)(4)(B)(ii).

[18] Secs. 1381 1383.

[19] Sec. 1382.

[20] Sec. 48.

[21] Sec. 38(b)(1).

[22] Sec. 39.

[23] See section II.A.3 of this pamphlet for a discussion of sec. 45 qualified facilities eligible for this provision.

[24] Sec. 54(f).

[25] Sec. 54D(a).

[26] Sec. 54(b).

[27] Sec. 54C(c); *see also* Notice 2009-33, 2009-17 I.R.B. (April 6, 2009) (soliciting applications for the present total national bond volume limitation for New CREBs and providing guidance on (1) eligibility requirements; (2) application requirements; (3) the method for allocating volume cap; and (4) certain other aspects of the rules).
[28] Sec. 54C(b).
[29] Sec. 54A(f).
[30] Sec. 54(d).
[31] Sec. 54(j)(4).
[32] Sec. 54(j)(5).
[33] Internal Revenue Service, Notice 2006-7, Clean Renewable Energy Bonds, 2006-1 C.B. 559 (March 6, 2006).
[34] Sec. 54(b).
[35] Sec. 54(e).
[36] Sec. 54(i).
[37] Sec. 54(h).
[38] This table is based on Announcement 2006-181 (November 20, 2006) and Announcement 2008-16 (February 8, 2008). Pursuant to the Energy Tax Incentives Act of 2005 (Pub. L. No. 109-58), the Treasury Department allocated $800 million of CREBs based on a "smallest-to-largest" project amount methodology -- beginning with the project requesting the smallest dollar amount and proceeding thereafter to projects for successively larger dollar amounts until the total national volume cap of $800 million was allocated among 610 projects (some of the allocations were subsequently relinquished). The allocations for governmental borrowers ranged from $23,000 to approximately $3.2 million. The allocations for cooperatives ranged from $120,548 to $31 million. The Tax Relief and Health Care Act of 2006 (Pub. L. No. 109-432) increased the available national allocation from $800 million to $1.2 billion. In 2008, the Treasury Department made a second round of allocations of approximately $406 million distributed among another 312 projects. For the second round, the allocations for governmental borrowers ranged from $15,000 to $2.95 million, and $300,000 to $30 million for cooperative borrowers.
[39] Sec. 54C(c); *see also* Notice 2009-33. Section 1111 of Title I of Division B of the American Recovery and Reinvestment Act of 2009 (Pub. L. No. 111-5) increased the national bond volume limitation for New CREBs by $1.6 billion, from $800 million to $2.4 billion.
[40] Sec. 54A(d)(2).
[41] Sec. 54A(d)(4).
[42] Given the differences in credit quality and other characteristics of individual issuers, the Secretary cannot set credit rates in a manner that will allow each issuer to issue tax credit bonds at par.
[43] See Internal Revenue Service, Notice 2009-15, *Credit Rates on Tax Credit Bonds*, 2009-6 I.R.B. 1 (January 22, 2009).
[44] For certain property, including tangible property used predominantly outside the United States, tax-exempt use property, tax-exempt bond-financed property, and certain other property, the MACRS "alternative depreciation system" of section 168(g) applies, generally increasing recovery periods and requiring straight-line depreciation.
[45] Sec. 168(e)(3)(B)(vi)(I).
[46] Only with respect to periods ending before January 1, 2017.
[47] Geothermal deposit is defined by reference to section 613(e)(2) as a geothermal reservoir consisting of natural heat which is stored in rocks or in an aqueous liquid or vapor (whether or not under pressure).
[48] Qualified fuel cell property is defined as a fuel cell power plant which has a nameplate capacity of at least 0.5 kilowatt of electricity using an electrochemical process, and has an electricity-only generation efficiency greater than 30 percent.
[49] Qualified microturbine property is defined as a stationary microturbine power plant which has a nameplate capacity of less than 2,000 kilowatts, and has an electricity-only generation efficiency of not less than 26 percent at International Standard Organization conditions.
[50] Sec. 168(e)(3)(B)(vi)(II).
[51] This definition is included by reference to section 48(l)(1 5) as in effect on the day before the date of the enactment (Nov. 5, 1990) of the Revenue Reconciliation Act of 1990.
The term "qualified fuel" means any synthetic solid fuel, and alcohol for fuel purposes if the primary source of energy for the facility producing the alcohol is not oil or natural gas or a product of oil or natural gas.
[52] The term "pollution control equipment" does not include any equipment which is installed on or in connection with property which, as of October 1, 1978, was using coal (including lignite), and was required to be installed by Federal, State, or local regulations in effect on such date.
[53] This equipment also includes equipment used for the storage of fuel derived from garbage at the site at which such fuel was produced from garbage.
[54] This definition is included by reference to the Federal Power Act (16 U.S.C. 796(17)(C)), as in effect on September 1, 1986.
[55] This definition is included by reference to the Federal Power Act (16 U.S.C. 796(17)(A)), as in effect on September 1, 1986.

[56] With the exception of biomass fuel property, property placed in service after December 31, 2008 and prior to February 17, 2009 qualifies for the new 30 percent credit rate (and $1,500 aggregate cap) if it met the efficiency standards of prior law for property placed in service during 2009. Biomass fuel property placed in service at any point in 2009 is governed by the new efficiency standard.

[57] This reference to the 2000 International Energy Conservation Code is superseded by the additional requirements described in the paragraph below regarding building envelope components.

[58] These standards are a seasonal energy efficiency ratio ("SEER") greater than or equal to 15, an energy efficiency ratio ("EER") greater than or equal to 12.5, and heating seasonal performance factor ("HSPF ") greater than or equal to 8.5 for split heat pumps, and SEER greater than or equal to 14, EER greater than or equal to 12, and HSPF greater than or equal to 8.0 for packaged heat pumps.

[59] These standards are a SEER greater than or equal to 16 and EER greater than or equal to 13 for split systems, and SEER greater than or equal to 14 and EER greater than or equal to 12 for packaged systems.

[60] Sec. 30B.

[61] See discussion surrounding Table 7, below.

[62] For hybrid passenger vehicles and light trucks, the term "maximum available power" means the maximum power available from the rechargeable energy storage system, during a standard 10 second pulse power or equivalent test, divided by such maximum power and the SAE net power of the heat engine. Sec. 30B(d)(3)(C)(i).

[63] In the case of such heavy-duty hybrid motor vehicles, the percentage of maximum available power is computed by dividing the maximum power available from the rechargeable energy storage system during a standard 10-second pulse power test, divided by the vehicle's total traction power. A vehicle's total traction power is the sum of the peak power from the rechargeable energy storage system and the heat (e.g., internal combustion or diesel) engine's peak power. If the rechargeable energy storage system is the sole means by which the vehicle can be driven, then the total traction power is the peak power of the rechargeable energy storage system.

[64] Sec. 54A(d)(1)(C).

[65] The American Recovery and Reinvestment Act of 2009 increased the national volume cap to $3.2 billion from $800 million.

[66] Sec. 54D(a). In addition, eligible issuer also include entities empowered to issue bonds on behalf of a State or local government and otherwise eligible issuers in conduit financing issues (as defined in Treas. Reg. sec. 1.150-1(b)). *See* Notice 2009-29, 2009-17 I.R.B. (April 6, 2009).

[67] *Id.*

[68] Sec. 54A(f).

[69] Sec. 54D(f)(1)(A). The American Recovery and Reinvestment Act of 2009 clarified that capital expenditures to implement green community programs include grants, loans and other repayment mechanisms to implement such programs. This expansion enables States to issue QEC bonds to finance, for example, retrofits of existing private buildings through loans and/or grants to individual homeowners or businesses, or through other repayment mechanisms. Other repayment mechanisms can include periodic fees assessed on a government bill or utility bill that approximates the energy savings of energy efficiency or conservation retrofits. Retrofits can include heating, cooling, lighting, water-saving, storm water-reducing, or other efficiency measures.

[70] Sec. 54D(f)(1)(B).

[71] Sec. 54D(f)(1)(C).

[72] Sec. 54D(f)(1)(D).

[73] Sec. 54D(f)(1)(E).

[74] Sec. 54D(d).

[75] *See generally*, Sec. 54D(e). *See also* Notice 2009-29 for 2009 (listing allocations to States of the national QEC bond volume cap).

[76] Sec. 54D(f)(2).

[77] Sec. 54D(h).

[78] Sec. 54A(d)(2).

[79] Sec. 54A(d)(4).

[80] Sec. 54A(d)(5).

[81] Sec. 54D(b) and Sec. 54A(b).

[82] Given the differences in credit quality and other characteristics of individual issuers, the Secretary cannot set credit rates in a manner that will allow each issuer to issue tax credit bonds at par.

[83] See Internal Revenue Service, Notice 2009-15, *Credit Rates on Tax Credit Bonds*, 2009-6 IRB 1 (January 22, 2009).

[84] See IRS notice 2006-52 and 2008-40.

[85] IRS Notice 2008-40 has set a target of a 10 percent reduction in total energy and power costs with respect to the building envelope, and 20 percent each with respect to the interior lighting system and the heating, cooling, ventilation and hot water systems.

[86] Sec. 43.

[87] Secs. 43(b) and 45K(d)(2)(C).

[88] Sec. 38(b)(8).

[89] An operator or working interest owner is defined as a person that holds a working or operating interest in any tract or parcel of land either as a fee owner or under a lease or any other form of contract granting working or operating rights.

[90] Sec. 263(c).

[91] Treas. Reg. sec. 1.612-4(a).

[92] Treas. Reg. sec. 1.612-4(b)(4).

[93] Sec. 291(b)(1)(A). The IRS has ruled that, if a company that has capitalized and begun to amortize IDCs over a 60-month period pursuant to section 291 ceases to be an integrated oil company, it may not immediately write off the unamortized portion of the capitalized IDCs, but instead must continue to amortize the IDCs so capitalized over the 60-month amortization period. Rev. Rul. 93-26, 1993-1 C.B. 50.

[94] Sec. 59(e)(1).

[95] In the case of IDCs paid or incurred with respect to an oil or gas well located outside of the United States, the costs, at the election of the taxpayer, are either (1) included in adjusted basis for purposes of computing the amount of any deduction allowable for cost depletion or (2) capitalized and amortized ratably over a 10-year period beginning with the taxable year such costs were paid or incurred (sec. 263(i)).

[96] The term "United States" for this purpose includes the seabed and subsoil of those submarine areas that are adjacent to the territorial waters of the United States and over which the United States has exclusive rights, in accordance with international law, with respect to the exploration and exploitation of natural resources (i.e., the Continental Shelf area) (sec. 638).

[97] Sec. 263A(c)(3).

[98] Sec. 193.

[99] Treas. Reg. sec. 1.193-1.

[100] Sec. 193(c).

[101] Sec. 193(b). A tertiary recovery method is any of the nine methods described in section 21 2.78(c)(1) - (9) of the June 1979 energy regulations, as defined in former section 4996(b)(8)(C), or any other method approved by the Internal Revenue Service.

[102] Sec. 193(b)(2).

[103] Sec. 148.

[104] Treas. Reg. sec. 1.148-1(e)(2)(iii).

[105] This amortization rule applies to G&G costs incurred in taxable years beginning after August 8, 2005, the date of enactment of the Energy Policy Act of 2005, Pub. L. No. 109-58. Prior to the effective date, G&G costs associated with productive properties were generally deductible over the life of such properties, and G&G costs associated with abandoned properties were generally deductible in the year of abandonment.

[106] Prior to the enactment of the Energy Independence and Security Act of 2007, major integrated oil companies were required to amortize G&G costs paid or incurred after May 17, 2006 over five years, as provided in Energy Tax Incentives Act of 2005.

[107] Generally, an integrated oil company is a producer of crude oil that engages in the refining or retail sale of petroleum products in excess of certain threshold amounts.

[108] Sec. 469.

[109] In the context of mineral extraction, depreciable assets are generally used to recover depletable assets. For example, natural gas gathering lines, used to collect and deliver natural gas, have a class life of 14 years and a depreciation recovery period of seven years.

[110] Secs. 611 - 613.

[111] Sec. 613A(c). The Code generally limits the percentage depletion method for oil and gas properties to independent producers and royalty owners.

[112] Sec. 613(a). For marginal production, discussed infra, this limitation is suspended for taxable years beginning after December 31, 1997, and before January 1, 2008, and for taxable years beginning after December 31, 2008 and before January 1, 2010.

[113] Sec. 613A(d)(1).

[114] Sec. 613(b)(4).

[115] Sec. 613(b)(1)(2)(B).

[116] Sec. 291(a)(2).

[117] Sec. 613(a).

[118] Sec. 613A(d)(2).

[119] Sec. 613A(d)(4).

[120] Sec. 613A(c).

[121] Sec. 613A(c)(8).

[122] Sec. 613A(c)(6).

[123] Sec. 613A(c)(6)(D).

[124] Sec. 613A(c)(6)(E).

[125] 1987-2 C.B. 674 (as clarified and modified by Rev. Proc. 88-22, 1988-1 C.B. 785).

[126] Under Code section 45K(c), qualified fuels are (1) oil produced from shale and tar sands; (2) gas produced from geopressured brine, Devonian shale, coal seams, or a tight formation, or biomass; and (3) liquid, gaseous, or solid synthetic fuels produced from coal (including lignite).

[127] This requirement also may be met by placing the property in service before January 1, 2010 or, in the case of self-constructed property, by beginning construction after June 14, 2005 and before January 1, 2008.

[128] The inflation adjustment is calculated using 2004 as the base year.

[129] For certain property, including tangible property used predominantly outside the United States, tax-exempt use property, tax-exempt bond-financed property, and certain other property, the MACRS "alternative depreciation system" of section 168(g) applies, generally increasing recovery periods and requiring straight-line depreciation.

[130] Sec. 168(i)(16).

[131] In the case of self-constructed property, the provision does not apply to property under construction on or before April 11, 2005.

[132] Sec. 168(i)(17).

[133] Generally, sec. 1245 property includes personal property. For a complete definition, *see* sec. 1 245(a)(3).

[134] Different rules apply with respect to NOLs arising in certain circumstances. For example, a three-year carryback applies with respect to NOLs (1) arising from casualty or theft losses of individuals, or (2) attributable to Presidentially declared disasters for taxpayers engaged in a farming business or a small business. A five-year carryback period applies to NOLs from a farming loss (regardless of whether the loss was incurred in a Presidentially declared disaster area). Special rules also apply to real estate investment trusts (no carryback), specified liability losses (10-year carryback), and excess interest losses (no carryback to any year preceding a corporate equity reduction transaction).

[135] Sec. 451(i).

[136] For example, a regional transmission organization, an independent system operator, or an independent transmission company.

[137] Sec. 38(b)(8).

[138] The 1.8-cents credit amount is reduced, but not below zero, if the annual average contract price per kilowatt-hour of electricity generated from advanced nuclear power facilities in the preceding year exceeds eight cents per kilowatt-hour. The eight-cent price comparison level is indexed for inflation after 1992.

[139] Sec. 45Q.

[140] Sec. 638(1).

[141] Sec. 638(2).

[142] For advanced coal project certification applications submitted after October 2, 2006, an electric generation unit using advanced coal-based generation technology designed to use subbituminous coal can meet the performance requirement relating to the removal of sulfur dioxide if it is designed either to remove 99 percent of the sulfur dioxide or to achieve an emission limit of 0.04 pounds of sulfur dioxide per million British thermal units on a 30-day average.

[143] The Secretary issued guidance establishing the certification program on February 21, 2006 (IRS Notice 2006-24).

[144] The second round of credit allocations were authorized on October 3, 2008, the date of enactment of the Energy Improvement and Extension Act of 2008 (Pub. Law 110-343, Div. B).

[145] The Secretary issued guidance establishing the certification program on February 21, 2006 (IRS Notice 2006-25).

[146] The second round of credit allocations were authorized on October 3, 2008, the date of enactment of the Energy Improvement and Extension Act of 2008 (Pub. Law 110-343, Div. B).

[147] Sec. 169.

[148] The amount attributable to the first 15 years is equal to an amount which bears the same ratio to the portion of the adjusted basis of the facility, which would be eligible for amortization but for the application of this rule, as 15 bears to the number of years of useful life of the facility.

[149] Sec. 291(a)(4).

[150] This provision was added by the Energy Policy Act of 2005 (Pub. L. No. 109-58) and is generally applicable to property that is constructed or acquired after April 11, 2005.

[151] Sec. 41(h).

[152] Under a special rule, 75 percent of amounts paid to a research consortium for qualified research are treated as qualified research expenses eligible for the research credit (rather than 65 percent under the general rule under section 41(b)(3) governing contract research expenses) if (1) such research consortium is a tax-exempt organization that is described in section 501(c)(3) (other than a private foundation) or section 501(c)(6) and is organized and operated primarily to conduct scientific research, and (2) such qualified research is conducted by the consortium on behalf of the taxpayer and one or more persons not related to the taxpayer. Sec. 41(b)(3)(C).

[153] Sec. 41(d)(3).

[154] Sec. 41(d)(4).

[155] Taxpayers may elect 10-year amortization of certain research expenditures allowable as a deduction under section 174(a). Secs. 174(f)(2) and 59(e).

[156] Sec. 280C(c).

[157] Sec. 280C(c)(3).

[158] Sec. 103(a).

[159] Secs. 141(b) and (c).

[160] The 10-percent private business use and payment threshold is reduced to five percent for private business uses that are unrelated to a governmental purpose also being financed with proceeds of the bond issue. In addition, as described more fully below, the 10-percent private business use and private payment thresholds are phased-down for larger bond issues for the financing of certain "output" facilities. The term output facility includes electric generation, transmission, and distribution facilities.

[161] See Treas. Reg. sec. 1.141-3(b)(4) and Rev. Proc. 97-13, 1997-1 C.B. 632.

[162] Sec. 141(b)(4).

[163] Sec. 141(d).

[164] Sec. 7704.

[165] Sec. 851.

[166] A British thermal unit is the amount of heat required to raise the temperature of one pound of water 1 degree Fahrenheit.

[167] This type of budgetary inefficiency can sometimes be tempered by targeting the credit at investment or expenditures above a base amount.

[168] In the conservation literature, this phenomenon of greater energy efficiency leading to behavioral responses that tend to increase the use of the more energy efficient equipment has come to be termed the "rebound effect," and has been estimated to reduce expected energy savings by up to 30 percent in the case of space heating and automobiles (see Frank Gottron, "Energy Efficiency and the Rebound Effect: Does Increasing Efficiency Decrease Demand?," Congressional Research Service report RS20981, July 2001.)

[169] This discussion assumes that the benefits across all types of alternative energy are equivalent, and assumed to be the displacement of a fossil fuel. In reality, different alternative energy sources might displace different types of fossil fuels, whose negative externalities may vary. Also, the production of certain renewables, such as solar or wind energy, may be more benign than the production of others, such as ethanol. Thus, depending on these other factors, varying credit rates could be economically efficient if there are differences across the renewable in the net benefits from each renewable and the fossil fuel it displaces.

[170] The sec. 45 electricity production credit is allowed only on the wind produced for the first ten years the facility is placed in service. If the existence of the credit induces a wind facility to be built that would not otherwise have been, and such a facility lasts for 20 years, then half of the wind produced from such facility does not receive any federal credit, and the true cost of the federal credit for that facility is half of what is shown on the table.

[171] The calculations were made by the Congressional Research Service. The comparison assumes both the hybrid motor vehicle and the comparable non-hybrid vehicle would have a life of 15 and one quarter years. Miles driven were not assumed to be constant across the 15.25-year life of the vehicles, but rather varied with the age of the vehicle. The comparable non-hybrid vehicles were the 2007 model Toyota Camry LE four-door sedan, the 2006 Honda Accord V6 EX four-door sedan, and the 2006 GMC Sierra K 1500 four wheel drive pickup truck.

[172] Severin Borenstein, "The Market Value and Cost of Solar Photovoltaic Electricity Production," Center for the Study of Energy Markets Working Paper 176, University of California, Berkeley, January 2008.

[173] If measured in terms of displaced fossil fuel consumption as was done in column 3 of Table 10, the comparable figure would be $3.33 per MMBtu of displaced fossil fuel consumption.

[174] If measured in terms of displaced fossil fuel consumption as was done in column 3 of Table 10, the comparable figure would be $5.32 per MMBtu of displaced fossil fuel consumption.

In: The Role of Taxes in Energy Production and Conservation ISBN: 978-1-61728-886-9
Editor: Bernadette M. Horton

Chapter 3

STATEMENT OF SENATOR MAX BAUCUS, CHAIRMAN, BEFORE THE COMMITTEE ON FINANCE, HEARING "TECHNOLOGY NEUTRALITY IN ENERGY TAX: ISSUES AND OPTIONS"*

Justice Oliver Wendell Holmes wrote: "[T]he best test of truth is the power of the thought to get itself accepted in the competition of the market. . . ."

Today we consider a new thought in the marketplace of ideas about tax incentives. The thought is that, in the creation of energy tax incentives, the government might not pick and choose among different technologies.

The thought is: The government might just set a performance standard, regardless of the technology employed. We could encourage things like reduction in greenhouse gas emissions, improvement in efficiency, or increased energy content. And then we would leave the job of picking the best technology to the competition of the market.

Folks call this sort of incentive structure "technology neutral."

There are several reasons why a tech-neutral approach to energy tax incentives might make sense.

First, it might well provide more bang for our energy-tax buck. By tying receipt of these credits to a common standard, we may be able to set the level of incentives more efficiently.

Second — and this may come as a surprise to many in the room sometimes government gets it wrong.

Consider the credit that Congress enacted in 1980 to stimulate oil shale, tar sands, and synthetic fuels from coal. The idea sounded good at the time. But many companies exploited the credit. Some sprayed coal with chemicals for no reason other than to line their pockets.

Third, a technology-neutral approach might mean that we have to change the tax code less often. Technology is bound to change faster than Congress can act. So there's appeal to instituting a series of incentives that we don't have to update all the time.

We've already taken some steps toward technology neutrality in recent legislation.

* This is an edited, reformatted and augmented edition of a statement/testimony, given before the United States Senate Committee on Finance, dated April 23, 2009.

For example, as part of last year's farm bill, Congress enacted the first-ever credit for cellulosic biofuels. Now this might sound like credit that picks a specific technology. But in fact, we passed a technology-neutral credit for cellulosic biofuels.

Another example is the approach that we took on coal incentives in last year's energy-tax bill. We removed the bias toward integrated gasification combined cycle facilities. And we put in its place a requirement that all recipients of clean coal tax credits meet at least a 65 percent standard for capture and storage of carbon dioxide.

This approach sounds sensible. Congress should not pick winners and losers. We should set a level playing field of standards for energy tax incentives. And we should let the marketplace foster competition.

But we need to beware of pitfalls.

For example, a couple years ago, Congress modified the alternative fuels tax credit. That's a 50- cent-a-gallon credit for a range of alternative fuels, including liquefied petroleum gas, compressed natural gas, liquid coal, and biomass-based fuel.

In modifying the definition of biomass-based fuel, the credit was inadvertently opened to apply to what is called "black liquor." Black liquor is a byproduct of the pulp-making process that has been used to power paper mills since the 1930s.

Paper companies learned that they could benefit from the alternative fuels credit by mixing a small amount of diesel with their black liquor, and then registering with IRS.

Unless we plug this loophole, the Federal Government is liable for billions in credits for black liquor in 2009 alone, even though the credit was never intended for this fuel.

So in this case, a more technology-neutral approach led to a dramatic spike in the use of the credit for an unforeseen purpose.

We are working to undo that unintended consequence. But our experience with black liquor suggests that we should exercise caution as we consider a tech-neutral approach. We have to make sure that we write the incentives correctly.

This committee has done a lot on energy tax incentives in recent months. And I'm proud of what we've achieved. But as we prepare for the next energy debate, including climate-change legislation, it may be time to consider an alternative means of promoting alternative energy.

And so, let us consider a new thought in the marketplace of ideas about tax incentives. Let us see if there is sense in getting the government out of the business of picking and choosing among different technologies. And let us see if technology-neutral incentives might just be a thought that gets accepted in the marketplace of ideas.

In: The Role of Taxes in Energy Production and Conservation ISBN: 978-1-61728-886-9
Editor: Bernadette M. Horton

Chapter 4

STATEMENT OF SEN. CHUCK GRASSLEY, RANKING MEMBER, U.S. SENATE, BEFORE THE COMMITTEE ON FINANCE, HEARING "TECHNOLOGY NEUTRALITY IN ENERGY TAX: ISSUES AND OPTIONS"*

As hockey and soccer players know, it's important to have goals. As we look at whether it makes sense to design technology-neutral energy tax incentives, we first need to consider what goals our energy tax policies seek to achieve. Some of the goals that have been mentioned for energy tax policy are a reduction in dependence on foreign oil, a reduction in the use of fossil fuels, and a reduction in carbon dioxide emissions. Depending on what goal or goals are selected, vastly different results emerge. For instance, if the only goal in the fuels arena is a reduction in dependence on foreign oil, then energy tax incentives that encourage more domestic drilling and oil production are appropriate. However, if the goal is solely a reduction in carbon dioxide emissions or a reduction in the use of fossil fuels, then those same energy tax incentives to encourage more domestic drilling and oil production are inappropriate.

Simplifying the energy-tax incentives by creating technology-neutral tax incentives is a noble ambition. However, getting consensus on what goals should be used in developing energy tax incentives can be a little like herding cats. Even if lawmakers agree on what goals should be used, which is a big "ifÓ, controversial issues arise. For example, whether nuclear energy should qualify for technology-neutral energy tax incentives would certainly be a controversial issue.

Also, the energy tax incentives that the Finance Committee has developed over the years have been extremely successful. For instance, the wind industry in the United States has made great advances with the help of the production tax credit, which I first authored in the early nineties.

Similarly, VEETC has helped the ethanol industry to reduce our dependence on foreign oil, improve our national security, and reduce carbon-dioxide emissions.

* This is an edited, reformatted and augmented edition of a statement/testimony, given before the United States Senate Committee on Finance, dated April 23, 2009.

As we move forward in designing energy-tax incentives, we need to be careful not to undo all the good work that this committee has done. Even the proponents of technology-neutral tax incentives acknowledge that certain technologies need more assistance in their early stages of development than others. They agree that this justifies a departure from technology-neutral energy tax incentives. I am interested in hearing the thoughts of this panel on these important energy-tax issues.

In: The Role of Taxes in Energy Production and Conservation ISBN: 978-1-61728-886-9
Editor: Bernadette M. Horton

Chapter 5

STATEMENT OF GILBERT E. METCALF, PROFESSOR ECONOMICS, TUFTS UNIVERSITY, MEDFORD, MA, BEFORE THE COMMITTEE ON FINANCE, HEARING "TECHNOLOGY NEUTRALITY IN ENERGY TAX: ISSUES AND OPTIONS"*

Chairman Baucus, Senator Grassley, and Members of the Committee, thank you for the invitation to testify this morning on the issue of technology neutrality in the treatment of energy in the tax system. I make the following points in my testimony today.

- Energy policy is shaped in important ways by the federal tax system. While taxes are one instrument of tax policy, subsidies in the form of accelerated depreciation, percentage depletion, production tax credits and investment tax credits are more commonly used instruments in the tax code.
- Technology neutrality can be defined in a variety of ways. It can be defined in terms of the effective tax rate on new investments in the sector, in terms of the levelized cost of power from new investments or in terms of specific policy goals that motivate energy tax incentives.
- Efficiency is best achieved by setting taxes on energy sources that have negative externalities associated with their production or consumption. Similarly technological neutrality is most easily achieved through the use of taxes.
- A second-best technological neutrality can be achieved through the use of subsidies but it is more difficult to do so. In particular it is very difficult to level the playing field across different non-polluting energy sources through the use of subsidies.

---

* This is an edited, reformatted and augmented edition of a statement/testimony, given before the United States Senate Committee on Finance, dated April 23, 2009.

I. BACKGROUND

Federal taxes specifically related to energy production or consumption are dominated by the federal motor fuels excise tax for the Highway Trust Fund. This 1 8.3¢ per gallon tax collected just under $40 billion in Fiscal Year 2006. In contrast taxes on coal to fund the Black Lung Disability Trust Fund collected $639 million in FY 2006 and the Leaking Underground Storage Tank tax collected $226 million in that year.[1]

The tax code has become an important instrument for energy policy over the past decade. Tax provisions for accelerated depreciation, percentage depletion, deductions and tax credits are different tools for reducing the cost of producing energy. The Energy Information Administration recently released a report detailing federal financial interventions in energy markets and notes that expenditures through the tax system account for nearly two-thirds of all federal support (see Table 1 below).[2]

Table 1. Federal Support for Energy: FY 2007 ($ Millions)

Fuel	Tax Expenditures	Total	Share of Total
Coal*	2,660	3,302	81%
Natural Gas and Petroleum Liquids	2,090	2,149	97%
Nuclear	199	1,267	16%
Renewable Energy	3,970	4,875	81%
Electricity (not fuel specific)	735	1,235	60%
End Use and Conservation	790	3,754	21%
Total	10,444	16,582	63%

Source: Table ES-1, Energy Information Administration. 2008. *Federal Financial Interventions and Subsidies in Energy Markets 2007*. Washington, DC: EIA SR/CNEAF/2008-01.

* - The 2007 tax expenditure for coal includes the credit for producing fuels from a non-conventional source in the amount of $2,370 million. Subsequent legislation has eliminated this tax expenditure for coal.

Table 2. US Oil Dependence

	1990	2000	2007	2030
Net oil imports as percent of total US Supply	42.2	52.9	58.2	55.5
World Oil Price (2007 $/BBL)	38	35	72	60
World Crude Production (million BBD)	65.5	74.9	81.5	102.9
OPEC Share (percent)	38.3	42.9	43.2	46.4
US Petroleum Consumption (million BBD)	17	19.7	20.7	22.8
US Share of World Production (percent)	26.0	26.3	25.4	22.2
Oil Intensity (1,000 BTUs/GDP) $2000	4.7	3.9	3.4	2.2
Oil Intensity (Value of oil as a percent of GDP)	2.6	2.0	3.6	1.9

Source: BP Statistical Review of World Energy (2008), EIA Annual Energy Review (2008), EIA Annual Energy Outlook (2008), EIA International Energy Outlook (2008)

Subsidies through the tax code play an especially important role in supporting fossil fuel and renewable energy production. They play a smaller role in supporting nuclear power production though this could change over the next decade. Production tax credits for new nuclear power production put in place in the Energy Policy Act of 2005 could significantly increase federal tax expenditures for this source of electricity.

The role of tax policy has increased significantly over the past decade. EIA documents that total federal subsidies and support for energy have roughly doubled between 1999 and 2007 (in year 2007 dollars). Over this period, tax expenditures have more than tripled from $3.2 billion in real terms to $10.4 billion.

As of 2007, EIA documented thirty seven tax expenditures related to energy production and consumption. The number of incentives in the tax code makes it difficult to assess their relative effectiveness and the extent to which they favor certain types of fuels over other fuels. I turn to this issue next.

But before doing so I wish to discuss *why* the federal tax system should intervene in energy markets through either taxes or subsidies. Economic theory provides clear prescriptions for situations where interventions through the tax code can improve social welfare. Externalities provide the most relevant rationale for the energy sector. If the production or consumption of energy has as a by-product the creation of an externality (e.g. pollution) then social welfare can be improved through government intervention. One way to do this is by taxing the externality. Thus a tax on the sulfur content of fossil fuels, for example, would be an efficient response to acid rain damages arising from fossil fuel consumption for electricity generation. This is an example of a Pigouvian tax.[3] It "internalizes the externality" by forcing firms to take into account the social costs of pollution by raising their private costs by the amount of the social damages that are generated by the pollutant. This approach implicitly makes clear that pollution generating activities have social benefits as well as costs. Optimal policy must balance those costs against the benefits; the tax is an efficient means of effecting that balance.

Rather than taxing activities that create negative externalities, we can provide subsidies to activities that are substitutes for externality generating activities. Put simply, if fuel X generates pollution damages while fuel Y does not, we can raise the price of fuel X relative to fuel Y to reflect the social damages from burning fuel X or we can reduce the price of fuel Y. Either approach encourages firms to use less of fuel X and more of fuel Y. This is the essential approach taken through federal energy tax policy. In large measure, we subsidize energy activities that we would like to encourage rather than tax activities that we would like to discourage.

What are the externalities that are of significant concern that drive federal tax policy towards energy? I would argue that two dominate the agenda. First is the concern with global climate change arising from increasing concentrations of greenhouse gases in the atmosphere. Fossil fuel combustion in the United States was responsible for eighty percent of domestic greenhouse gas emissions in 2007.[4] Any policy to reduce U.S. greenhouse gas emissions must have as a key element incentives to shift from fossil to renewable fuels consumption.

A second concern is our heavy reliance on petroleum products and the dominance of this fuel in the transportation sector. In 2007 seventy percent of petroleum products were used by the transportation sector. Conversely, petroleum accounted for over 95 percent of the fuel used in this sector. Our reliance on petroleum makes us vulnerable to economic dislocations from sharply rising oil prices or supply disruptions. Table 2 illustrates our increasing reliance on oil over the past few decades. Oil imports have risen from just over 40 percent of total US supply to nearly 60 percent in 2007. The EIA Annual Energy Outlook does not project any significant decline in this share over the next few decades under current policy. Many have argued that our heavy reliance on oil constrains our foreign policy, drives up our military costs, and makes us vulnerable to macroeconomic shocks when oil prices rise as they did over the past few years.[5]

Energy production and consumption are associated with negative externalities in addition to climate change and oil dependence. I do not focus on those here because many of these negative externalities are currently addressed through regulatory means. For example, the Acid Rain Program run by the Environmental Protection Agency has been a highly cost-effective response to the damages from releasing sulfur dioxide in fossil fuel electric generation units. Moreover the current set of energy subsidies is arguably focused to a large extent on reducing greenhouse gas emissions and reducing our consumption of oil. For the purposes of this testimony I will take as given that going forward tax policy will be predominantly concerned with these two issues and that any assessment of energy tax policy must consider, among other things, the degree to which policy reduces greenhouse gas emissions or our reliance on petroleum products.

II. What Does Technology Neutrality Mean?

This hearing is concerned with technology neutrality in energy production. Before assessing this concept we need to define it. In a general sense technology neutrality means that our tax code does not favor one fuel over another. With this as our definition our tax code is not technology neutral nor should it be. To the extent that certain energy sources create negative externalities we want to ensure that the tax code (or federal policy more generally) takes into account the pollution arising from energy production or consumption. We can modify the definition to mean that the tax code should not favor one fuel over another after taking into account any positive or negative externalities arising from the production or consumption of energy.

While conceptually straightforward, it is more difficult in practice to identify whether certain energy technologies are advantaged or disadvantaged by the tax code. We cannot observe what the mix of energy technologies and fuels would be in the absence of a technology neutral tax system. Moreover efforts to measure the impact of changes in the tax code on energy production and consumption are made more difficult by the fact that changes in energy tax provisions often occur at the same time as (or soon after) significant changes in energy prices or supply.

One approach to quantify the impact of the tax system on energy investment is the construction of *effective tax rates*. An effective tax rate is a summary measure of the various provisions in the tax code that affect investment in new capital. Specifically, it compares the before-tax return to the difference between the before- and after-tax return. The before-tax return is the return an investment must earn in order to cover its cost, pay the required return to investors, and pay taxes on the project. The after-tax return is the return that savers (the source of funds for investment) expect to receive after taxes are paid on marginal investments. Thus, if savers are prepared to accept seven percent on an investment after tax and the project must earn ten percent in order to cover depreciation, taxes, and required payments to investors, the effective tax rate is 30 percent $\left(\frac{10-7}{10}\right)$.

Effective tax rates focus on the marginal cost of funding investments rather than on project cost. In particular, they focus on the cost of a break-even investment. Because they summarize the many provisions of the tax code that affect the returns on capital investment, effective tax rates are frequently used to consider how the tax system affects capital investment. This is a particularly salient issue given the capital investment needs of energy infrastructure in the United States.

Table 3 reports estimates of effective tax rates on new energy investment assuming the tax rules in place in 2007.[6]

Table 3 illustrates that new energy capital investments for many fuels can have large and negative effective tax rates. An effective tax rate of -100 percent, for example, means that the return an investment must earn prior to paying taxes need only be half as large as the return investors require since the tax code will provide sufficiently generous tax treatment that the project return increases to the investor's required return. The table also illustrates that tax credits for certain new electricity generating units are the predominant source of the tax benefits for these technologies.[7]

Another way to report subsidies in the tax code is the subsides per BTU of energy or MWh of electricity generation. Table 4 reports data from the EIA study discussed above.

Table 3. Effective Tax Rates on New Energy Investment

	Current Law	**No Tax Credits**	**Economic Depreciation**
	(1)	(2)	(3)
1. Electric Utilities			
Generation			
Nuclear	-99.5%	32.4%	-49.4%
Coal (PC)	3 8.9%	3 8.9%	3 9.3%
Coal (IGCC)	-11.6%	38.9%	-10.3%
Gas	34.4%	34.4%	39.3%
Wind	-163.8%	12.8%	-13.7%
Solar Thermal	-244.7%	12.8%	-26.5%
2. Petroleum			
Oil Drilling (non-integrated firms)	-13.5%	-13.5%	39.3%
Oil Drilling (integrated firms)	15.2%	15.2%	39.3%
Refining	19.1%	19.1%	39.3%
3. Natural Gas			
Gathering Pipelines	15.4%	15.4%	39.3%
Other Pipelines	27.0%	27.0%	39.3%

Source: Table 2, Metcalf, Gilbert E. 2009. Taxing Energy in the United States: Which Fuels Does the Tax Code Favor? New York: The Manhattan Institute. PC stand for pulverized coal and IGCC for integrated gasification combined cycle.

Table 4. Subsidies per Unit of Energy Production in 2007

<table>
<tr><th></th><th>Energy</th><th>Electricity</th></tr>
<tr><td></td><td>$/billion BTUs</td><td>$/MWh</td></tr>
<tr><td>Coal</td><td rowspan="2">113</td><td>0.14</td></tr>
<tr><td>Refined Coal*</td><td>29.94</td></tr>
<tr><td>Natural Gas and Petroleum Liquids</td><td>63</td><td>0.22</td></tr>
<tr><td>Nuclear</td><td>24</td><td>0.25</td></tr>
<tr><td>Renewable Energy</td><td>584</td><td>2.01</td></tr>
</table>

Source: Energy Information Administration. 2008. *Federal Financial Interventions and Subsidies in Energy Markets 2007*. Washington, DC: EIA SR/CNEAF/2008-01.

* - The 2007 tax expenditure for coal includes the credit for producing fuels from a non-conventional source in the amount of $2,370 million. Subsequent legislation has eliminated this tax expenditure for coal.

The first column reports the total tax subsidy for energy per billion BTUs of production. I have combined refined coal and coal here given data availability. The subsidy for coal is roughly double that of natural gas and petroleum liquids and roughly five times that of the subsidy for nuclear power.[8] In contrast, the subsidy per billion BTUs of renewable energy is nearly $600. The second column restricts attention to subsidies for fuels used in the production of electricity. Here I've broken out refined coal given data on electricity generation with refined coal. The subsidy per megawatt hour of electricity production is highest for refined coal and lowest for other coal.

Measuring subsidies per dollar of production is problematic for a number of reasons. Table 4 measures the average subsidy but provides no information about the subsidy's effect on the use of this fuel. It may be that production of a particular form of energy would occur in the absence of any subsidy directed at that fuel source. Second, the subsidy doesn't take into account differences in the quality of fuels. On an energy content basis, natural gas is nearly five times the cost of coal. Thus while the subsidy to regular coal used in the production of electricity is roughly two-thirds that of natural gas on a MWh basis, the coal subsidy is more beneficial per dollar of spending on coal. Third, the subsidy is not related to any externality that may be driving energy policy. Whether the subsidy for renewable energy is high or low depends on the benefits that come about from the reduction in our use of that fuel. We cannot say anything about that by focusing on a subsidy per unit of energy.

A final way to measure subsidies is per ton of carbon dioxide emissions that is not emitted or barrel of oil that is not consumed. The benefit of this approach is that it calibrates the measure of the tax code's impact to the policy goals we care about (reducing greenhouse gas emissions and oil consumption). If the tax subsidy per ton of avoided greenhouse gas emissions from technology X is twice that of reducing emissions from technology Y then we can say that our tax policy favors technology X over Y on this dimension.

This definition of technology neutrality is not the same as efficiency in abatement of pollution. The latter requires that the marginal cost of pollution abatement be equalized across energy sources. Unless subsidies are designed in terms of a payment per unit of pollution reduced it is difficult if not impossible to achieve economic efficiency across fuel types. Moreover, as I discuss below, even if subsidies are constructed in this fashion, it is difficult to disentangle true emission reductions from reductions that would have taken place in the absence of the tax subsidy.

III. Achieving Technology Neutrality through a Subsidy Based Policy

Using subsidies within the tax system to achieve energy policy goals has been a time honored custom throughout the history of the U.S. income tax. It is important, however, to recognize the limitations of subsidies in achieving efficient outcomes. Congress may decide that the political benefit of a subsidy based approach outweighs the efficiency costs but it should be aware of the drawbacks of this approach so as to use the instrument as efficiently as possible.

First note that a subsidy based approach achieves the important goal of adjusting relative prices of polluting and non-polluting energy sources in the right direction. If fuel source X causes pollution that is equal to 10 percent of its cost then we can provide the right incentive to fuel users choosing between fuel sources X and Y by raising the price of X by 10 percent or by lowering the cost of fuel source Y by 1/(1.10) or 9.1 percent. Either way the relative cost of fuel source X to Y is now ten percent higher than it was prior to the implementation of new energy policy. Either a tax or a subsidy can be effective on the margin of choosing among fuel sources where some sources cause pollution.

This creates a problem, however, on a different margin. Efficiency requires that consumers make decisions taking into account the full cost of using commodities – including the pollution costs associated with using energy. Raising the cost of the polluting fuel source X raises the overall cost of energy use and encourages a reduction in energy consumption. More precisely, consumers shift away from consuming energy to consuming other goods. This substitution is driven by the higher overall cost of energy. Subsidizing the clean substitute undermines this consumer substitution effect as it leads to a lower cost of energy overall. Consumers do not reduce energy consumption as much as they would under a cost-raising policy.

Second, subsidies that appear to be technologically neutral may not be neutral at all in the sense of equalizing the subsidy cost per unit of activity that Congress is trying to discourage. Consider the tax credit for hybrid vehicles put in place in the Energy Policy Act of 2005. The credit ranges from zero to $3,000 per vehicle depending on whether the vehicle meets the specific hybrid criteria and on how many vehicles have been sold. The credit phases out as the vehicle hits certain sales targets over time. Table 5 shows the subsidy cost per gallon of gasoline saved through this credit for a number of vehicles. The tax credit is for model 2009 vehicles. I measure the savings relative to a vehicle that gets 20 miles per gallon assuming the vehicle is driven the average number of miles currently driven by private vehicles in the United States.

The table illustrates several points. First, the tax credit per gallon of gasoline saved varies from zero to over $11 per gallon. Second, certain hybrid vehicles that get high mileage are excluded from the credit because they have been successful in the market place. Third, certain high mileage vehicles are excluded from the subsidy because they do not use specified technology. Note that the Corolla gets nearly the same mileage as the Tribute Hybrid. This is the most egregious violation of technology neutrality. The tax credit provides no incentive to tinker with the internal combustion engine to achieve increases in vehicle efficiency despite the many opportunities that exist to make the internal combustion engine more efficient. Our tax policy should provide the same incentives to improve mileage regardless of the technology put in place. Only in this way is true technology neutrality achieved.

The hybrid vehicle tax credit is a clear example of inefficient allocation of resources across fuel saving capital investments. It is not the only example, however. Inefficient allocations can occur even when policies appear to be technology neutral. Consider the production tax credit for electricity generated from renewable sources. Currently the tax credit is worth 2.1¢ per kWh for electricity over the first ten years of the plant's life.[9] This policy appears to be technology neutral (assuming all renewable technologies are made eligible for the credit). Renewable in this context means carbon- free. But consider Table 6 which compares the production tax credit for wind with that for geothermal energy.

The subsidy per ton of carbon dioxide avoided critically depends on which power source is displaced by the new renewable capacity addition. Geothermal power, for example, has a capacity factor of over 70 percent – meaning that it is producing power on average for 70 percent of the year – while wind's capacity factor is less than 30 percent.[10] Geothermal power is more likely to displace base load coal units than natural gas while the opposite is true for wind. Under the assumption that geothermal displaces coal and wind displaces natural gas, the subsidy for the former is $7.74 per ton of carbon dioxide avoided while the subsidy for wind is $12.28 per ton. The difference arises because coal emits on average one ton of CO2 per MWh of electricity generation while natural gas emits on average roughly two-thirds of a ton of CO_2 per MWh.

The point here is not whether geothermal displaces coal and wind natural gas (or even whether the displaced fuel is constant over time). Rather the point is that a technology neutral policy focused on reducing greenhouse gas emissions should favor technologies that are more likely to displace coal than natural gas. The current new technology credits do not take this into account.

Table 5. Hybrid Vehicle Tax Credit Model 2009 Values

Vehicle	MPG	Hybrid Vehicle Tax Credit	Annualized Value of Credit	Annual Gasoline Savings (Gallons)	Tax Credit per Gallon of Gasoline Saved
Chrysler Aspen Hybrid	21	$2,200	$347	30	$11.68
Ford Escape Hybrid (2WD)	32	$3,000	$474	234	$2.02
Mazda Tribute Hybrid (2WD)	32	$3,000	$474	234	$2.02
Nissan Altima Hybrid	34	$2,350	$371	257	$1.44
Toyota Corolla	31	$0	$0	222	$0
Toyota Prius	46	$0	$0	353	$0

Source: Author's calculations of savings relative to a vehicle that gets 20 miles per gallon and is driven 12,485 miles per year. Vehicles are assumed to be driven for ten years and savings are annualized with a ten percent discount rate.

Table 6. Production Tax Credit

Renewable Source	PTC	Capacity Factor	Subsidy per on CO_2
Geothermal	$ 0.02 1	73%	$ 7.74
Wind	$ 0.021	27%	$ 12.28

Source: Author's calculations. Capacity factor based on electricity generation in 2006. CO_2 emissions avoided assume geothermal replaces coal fired base load capacity while wind replaces natural gas shoulder or peaking capacity. Coal and natural gas emissions based on EIA estimates

In summary, the current set of subsidies to encourage reductions in petroleum consumption and greenhouse gas emissions have two drawbacks. First, they generate a distortion on the margin between energy consumption and consumption of other non- energy commodities. Second, they generate distortions among the externality-reducing technologies in a way that raises the cost of achieving our policy goals.

IV. Design Issues

In addition to the pricing issues discussed above, the current set of energy tax initiatives have other issues that could fruitfully be addressed by lawmakers. The first issue is that of stability and clarity in the policy. The historic pattern of two-year authorization cycles for production tax credits has created great uncertainty in the wind industry and led to boom and bust cycles that raise the cost of renewable energy investment.[11] Greater certainty over the production tax credit would smooth out investment and reduce bottlenecks in turbine manufacture that delay projects and raise costs. A related issue is the ability to use tax benefits. One casualty of the current financial crisis is the reduced tax appetite of firms that historically have invested in wind and other renewable projects. The provision of a rebate option in the American Recovery and Reinvestment Act of 2009 addresses this concern.

A second key design issue is that of additionality. Does the policy lead to incremental reductions in pollution or simply subsidies for emission reducing activities that would have occurred in the absence of the policy? A good example of this is the $.50 per gallon alternative fuels mixture credit. This credit is intended to encourage the addition of biodiesel and other biomass based fuels to petroleum to reduce petroleum use. Recently it has emerged that many paper firms are taking the credit for mixing diesel fuel with black liquor, a biomass by-product of paper making that historically has been used by the industry as a fuel source for their boilers. Controversy has arisen over whether paper firms are adding diesel fuel to black liquor purely for the purpose of claiming the tax credit biodiesel mixture tax credit.[12] This is troubling on two levels. First, it may be highly inefficient if credits are being provided for inframarginal activities. This is a common problem with any subsidy. We want to provide the incentive to firms that would not have undertaken the desirable activity in the absence of the subsidy. But we don't want to provide the subsidy to firms that would have undertaken the activity regardless of the subsidy. But the example from the paper industry is troubling beyond the inframarginal nature of the subsidy. If the tax credit is raising the demand for diesel fuel in order to make the biofuel eligible for the credit, then it is having the perverse effect of raising rather than lowering demand for petroleum products.[13]

A third important design issue is the interaction between tax policy and other policies. A simple example here is the interaction of the hybrid vehicle tax credit and the Corporate Average Fuel Economy (CAFE) standards. Allowing tax credits for hybrids encourages the production and purchase of high mileage vehicles. But CAFE sets minimum fleet mileage standards for automakers. Producing more hybrid vehicles relaxes the CAFE mileage constraint for automakers and allows them to sell more low mileage vehicles.[14] One possible policy response to this would be to exclude credit receiving hybrids from the fleet for purposes of meeting CAFE standards. Alternatively one could eliminate the credit and simply let CAFE be the driving incentive for hybrid production.

V. A Better Approach

I have identified a number of problems with the current approach. Energy related tax subsidies lower rather than raise the cost of consuming energy. Much of the subsidy may be inframarginal. And the policy can be undermined through interaction with other energy policies. Here I wish to briefly mention policies that avoid most if not all of these pitfalls.

Assuming our concern is with climate change and oil consumption, optimal policies will raise the cost of emitting greenhouse gases and oil consumption.[15] One approach to discourage greenhouse gas emissions is through a carbon pricing mechanism.[16] One approach is through a carbon fee. Elsewhere I describe a proposal to price carbon emissions in a way that meets targets for emission caps over a control period (say from 2012 through 2050) to ensure that environmental goals are met while achieving price stability.[17]

A simple and efficient way to reduce oil consumption is to implement an oil consumption tax. Because of the volatility of oil prices and occasional spikes as we saw last year, I proposed (along with a colleague) a variable oil consumption tax that phases out as oil prices rise.[18] An oil consumption tax is preferable to an increase in the gasoline tax since it targets all oil consumption rather than the portion targeted to motor vehicles. But an increase in the gasoline tax in lieu of an oil consumption tax would go a long way towards improving efficiency.[19]

Both of these approaches address the problems addressed above. They ensure that energy consumption internalizes the costs of externalities associated with its production or consumption and achieves the socially efficient mix of energy and non- energy consumption. Second, they avoid problems of inframarginal subsidies or perverse incentives. Third, they complement rather than work at cross purposes with other federal energy policies.

VI. Conclusion

Current energy tax policy can perhaps be best viewed as a transitional policy until policies such as carbon pricing (whether through a carbon fee or a cap-and-trade system) are put in place along with consideration of an oil consumption tax or increase in the gas tax. In the meantime, Congress should consider how they might best modify the existing subsidies in the tax system to achieve true technology neutrality.

True technology neutrality requires measuring the subsidy cost of reducing the externality in question. Here I have focused on reducing greenhouse gas emissions and oil consumption. Policies should provide a level playing field in the sense that the subsidy per unit of externality avoided should be comparable across technologies. They should also consider the extent to which true reductions in the externality occur and avoid unintended consequences. This is all very easy to say but difficult to do. But so long as our energy policy is built around providing subsidies for activities we wish to support as opposed to taxing those activities we wish to discourage, we will always face difficult design problems that complicate our efforts to achieve efficient and cost effective outcomes.

Thank you for the opportunity to testify today.

Gilbert E. Metcalf is a Professor of Economics at Tufts University and a Research Associate at the National Bureau of Economic Research. He is also a Research Associate at the Joint Program on the Science and Policy of Global Change at MIT. Metcalf has taught at Princeton University and the Kennedy School of Government at Harvard University and been a visiting scholar at MIT.

Metcalf has served as a consultant to numerous organizations including, among others, the U.S. Department of the Treasury, the U.S. Department of Energy, and Argonne National Laboratory. He currently serves as a member of the National Academy of Sciences Committee on Health, Environmental, and Other External Costs and Benefits of Energy Production and Consumption. In addition he serves or has served on the editorial boards of *The Journal of Economic Perspectives*, *The American Economic Review*, and the *Berkeley Electronic Journals in Economic Analysis and Policy*.

Metcalf's primary research area is applied public finance with particular interests in taxation, energy, and environmental economics. His current research focuses on policy evaluation and design in the area of energy and climate change. He has published papers in numerous academic journals, has edited two books, and has contributed chapters to several books on tax policy. Metcalf received a B.A. in Mathematics from Amherst College, an M.S. in Agricultural and Resource Economics from the University of Massachusetts Amherst, and a Ph.D. in Economics from Harvard University.

End Notes

[1] Statistics taken from the Budget of the United States (2009), Historical Tables, Table 2.4. See Metcalf, Gilbert E. 2007. Federal Tax Policy towards Energy. *Tax Policy and the Economy* 21:145-184 for further discussion of the federal taxes on energy along with a comparison and contrast with other countries.

[2] Energy Information Administration. 2008. *Federal Financial Interventions and Subsidies in Energy Markets 2007*. Washington, DC: EIA SR/CNEAF/2008-01.

[3] Named for the economist Arthur C. Pigou, an early proponent of this policy instrument in Pigou, Arthur C. 1938. *The Economics of Welfare*. London: Weidenfeld and Nicolson. A comparable approach – and the one taken to address acid rain – is to create a cap-and-trade system for SO_2. Either approach puts a price on emissions of SO_2 and provides the appropriate price signal to electric utilities to reduce emissions.

[4] See Environmental Protection Agency. 2009. *Inventory of U.S. Greenhouse Gas Emissions and Sinks: 1990 - 2007*. Washington, DC: Environmental Protection Agency, EPA 430-R-09-004.

[5] On the first point, see Deutch, John, and James Schlesinger. 2006. National Security Consequences of U.S. Oil Dependency Washington, DC: Council on Foreign Relations Task Force Report No. 58. On the macroeconomic impact of oil shocks, see – among other sources – Hamilton, James. 2009. Causes and Consequences of the Oil Shock of 2007-2008. Washington, DC: Brookings Institution.

[6] This analysis comes from Metcalf, Gilbert E. 2009. *Taxing Energy in the United States: Which Fuels Does the Tax Code Favor?* New York: The Manhattan Institute.

[7] A similar approach that focuses on the cost of producing electricity is to report the levelized cost of a project. This is the constant revenue per kWh of electricity generation that a project must earn over its life to cover its costs. One can compare levelized cost measures with under different tax assumptions to see how the tax code affects the cost of a project. This is the approach taken in Metcalf, Gilbert E. 2007. Federal Tax Policy towards Energy. *Tax Policy and the Economy* 21:145-184

[8] Note that no developer has yet made use of the production tax credits for new nuclear power plants.

[9] Certain sources (e.g. municipal solid waste and open loop biomass) are eligible for a tax credit at half this rate.

[10] The capacity factor for wind depends importantly on location and turbine design. Capacity factors as high as 40 percent are not out of the question. But even at higher capacity factors the point of this example is unaffected.

[11] The American Recovery and Reinvestment Act of 2009 extends the production tax credit (PTC) for wind through 2012 and allows PTC qualified facilities to opt for a 30 percent investment tax credit or a cash rebate. These options are described in greater detail in Bolinger, Mark, Ryan Wiser, Karlynn Cory, and Ted James. 2009. *PTC, ITC, or Cash Grant?* Berkeley: Lawrence Berkeley National Laboratory LBNL-1642E.

[12] See Mouawad, Jad, and Clifford Krauss. 2009. Lawmakers May Limit Paper Mills' Windfall. *New York Times*, April 18, 2009.

[13] The perverse impact of policy is not limited to the biodiesel mixing tax credit. Research by Holland, Hughes, and Knittel suggest that low carbon fuel standards may have the perverse effect of increasing net carbon emissions. See Holland, Stephen P., Jonathan E. Hughes, and Christopher R. Knittel. 2009. Greenhouse Gas Reductions under Low Carbon Fuel Standards? *The American Economic Journal: Economic Policy* 1 (1):106-146.

[14] Alternatively and equivalently, it leads to the substitution of hybrid vehicles for other high mileage vehicles that in the absence of hybrids the automakers market primarily to meet CAFE fleet standards.

[15] Clearly there is overlap between policies that discourage oil consumption and greenhouse gas emissions. But policies can also work at cross purposes. A desire to reduce oil consumption could lead to increased coal consumption (and greenhouse gas emissions) if plug-in cars are a key part of the strategy to reduce oil consumption. Hence it is desirable to have multiple policy instruments in the face of multiple policy goals.

[16] I use the term carbon price as this is the common terminology despite the fact that the price can extend to gases beyond carbon dioxide.

[17] See Metcalf, Gilbert E. *Reacting to Greenhouse Gas Emissions: A Carbon Tax to Meet Emission Targets*, Tufts Department of Economics Working Paper 2009-03. For a detailed description on how to implement a carbon fee see Metcalf, Gilbert E., and David Weisbach. forthcoming. The Design of a Carbon Tax. *Harvard Environmental Law Review*.

[18] See Bordoff, Jason and Gilbert E. Metcalf, *Breaking The Boom-Bust Oil Cycle*, The New Republic Blog (The Vine), Jan. 6, 2009. Available at
http://blogs.

[19] Research finds that the optimal tax on gasoline in the United States falls far short of the unpriced social cost of its use. See Parry, Ian, and Kenneth A. Small. 2005. Does Britain or the United States Have the Right Gasoline Tax? *American Economic Review* 95:1276-1289.

In: The Role of Taxes in Energy Production and Conservation ISBN: 978-1-61728-886-9
Editor: Bernadette M. Horton

Chapter 6

TESTIMONY OF DR. DAVID L. GREENE, VISITING SCHOLAR, INSTITUTE FOR TRANSPORTATION STUDIES, UNIVERSITY OF CALIFORNIA AT DAVID, AND CORPORATE FELLOW AT OAK RIDE NATIONAL LABORATORY, BEFORE THE COMMITTEE ON FINANCE, HEARING "TECHNOLOGY-NEUTRAL INCENTIVES FOR ENERGY-EFFICIENT LOW GREENHOUSE GAS EMITTING VEHICLES"*

Good morning Mr. Chairman, Senators and distinguished guests. Thank you for the opportunity to offer my views on the pros and cons of technology-neutral energy and environmental incentives. I will confine my remarks to the area I know best, which is the transportation sector, addressing primarily incentives for energy efficient and low greenhouse gas (GHG) emitting vehicles.

For most energy and environmental policy goals, performance-based, technology-neutral incentives (and standards) are superior to those that target a specific technology. Performance- based incentives allow the widest scope for innovation, and permit market forces the greatest latitude to select and implement cost effective solutions. Because of our limited ability to foresee technological solutions that are possible but do not yet exist, it is almost always more effective and economical to specify the energy or environmental objective rather than a specific means of achieving it. Well designed fiscal incentives provide a clear and consistent signal to the market to make continuing progress toward energy goals. In certain cases they can even correct limitations of the marketplace.

Performance can be measured in different ways: e.g., energy use, petroleum consumption, or GHG emissions.[1] The choice could be important because some fuels that would help reduce petroleum dependence (e.g., coal-derived gasoline) would increase full fuel cycle GHG emissions unless by-product carbon dioxide (CO_2) were captured and

* This is an edited, reformatted and augmented edition of a statement/testimony, given before the United States Senate Committee on Finance, dated April 23, 2009.

sequestered. Choosing GHG emissions as a metric however, would substantially benefit energy security since the preponderance of measures for reducing GHG emissions from transportation vehicles will also reduce oil dependence.

Why Vehicle Incentives Can Help Achieve Energy Goals

The market system is the fundamental mechanism by which we will achieve our national energy goals. However, markets run into difficulties in several areas. Emissions of GHGs from the combustion of fossil fuels are a near-perfect example of a public good externality that requires public policy solutions. At the center of our oil security problem is the monopoly influence of the OPEC cartel. The nationally owned oil companies controlling four out of five barrels of the world's proved reserves and well over half of the world's ultimately recoverable resources of conventional oil create oil price shocks, inflate world oil prices and in the process appropriate hundreds of billions of dollars of wealth from oil consuming economies. By my estimates, oil dependence cost our economy between $700 and $800 billion dollars in 2008. The market problem here is not externalities but monopoly power, and fiscal policies alone are not likely to solve the problem (Leiby, 2007).

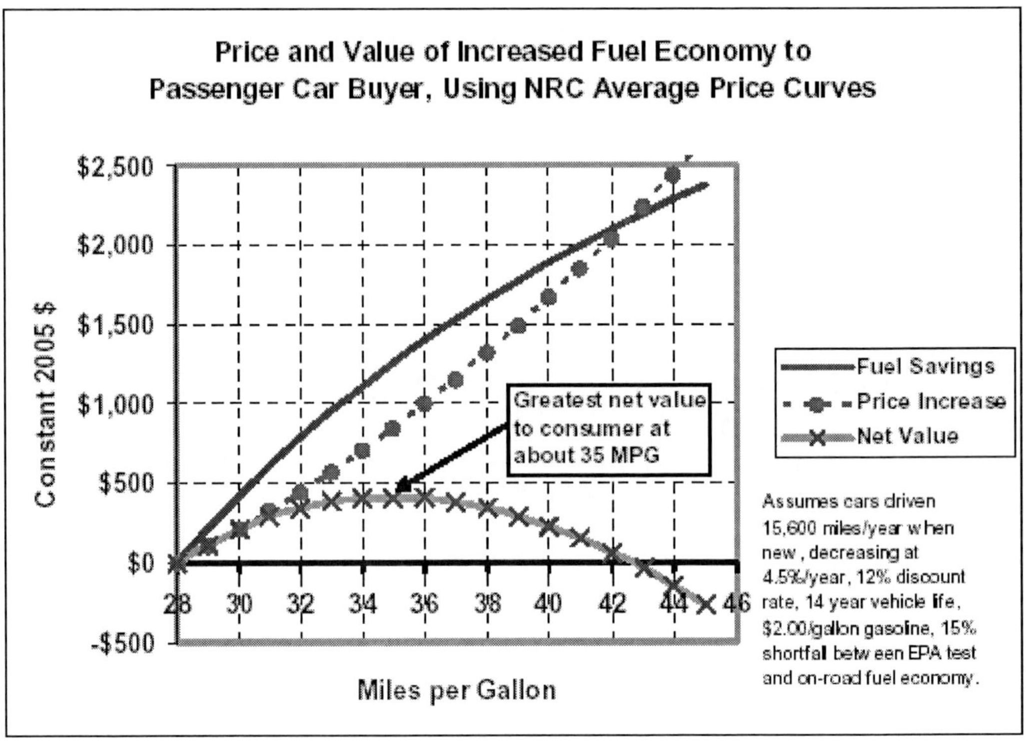

Figure 1. Illustrative Private Cost and Expected Benefit of Increasing Passenger Car Fuel Economy. (Greene, German and Delucchi, 2009)

Markets for energy efficiency in general, and the market for automotive fuel economy in particular, also have important limitations. As a general rule, more efficient automotive technologies cost more and deliver benefits in the form of future fuel savings. The estimated cost of increasing the fuel economy of an average U.S. passenger car based on the 2002 National Research Council (NRC) fuel economy study is illustrated by the red dotted line in Figure 1. The expected present value of future fuel savings is shown as a solid gray line. Of greatest interest to the consumer is the difference between the two, the expected net present value. This increases to a maximum of about $400 at 35 miles per gallon (MPG), the point at which the marginal cost of increasing fuel economy exactly equals the marginal value of expected fuel savings. Figure 1 reflects private costs only; motor fuel taxes are included but no values are attached to reducing GHG emissions or oil dependence.

Figure 1 represents future fuel savings as a known quantity. However, consumers view future fuel savings as uncertain due to ambiguity about future energy prices, the validity of official fuel economy estimates, vehicle life expectancies, future vehicle travel and other factors. When the uncertainty about future payoffs is considered, the net value of increasing fuel economy at each higher fuel economy level becomes a probability distribution, as shown in Figure 2. As Figure 2 shows, if fuel prices are low and the vehicle's rated fuel economy is not realized, the consumer might actually lose money despite the expected gain of $405.

Nobel prize-winning economic research conducted over the past three decades has established that, in general, consumers are loss-averse. That is, they weight potential losses from a risky bet more heavily than potential gains. When the inherent loss-aversion of typical consumers is taken into account, technologies that are cost-effective in terms of their expected payoff appear to be too risky. Using the same data and assumptions of the 2002 NRC study of the CAFE standards (NRC, 2002) but applying a typical loss-aversion function (Tversky and Kahneman, 1992) changes the perceived value of the fuel economy bet from an expected net gain of $405 to a perceived loss equal to -$32 (Figure 3).

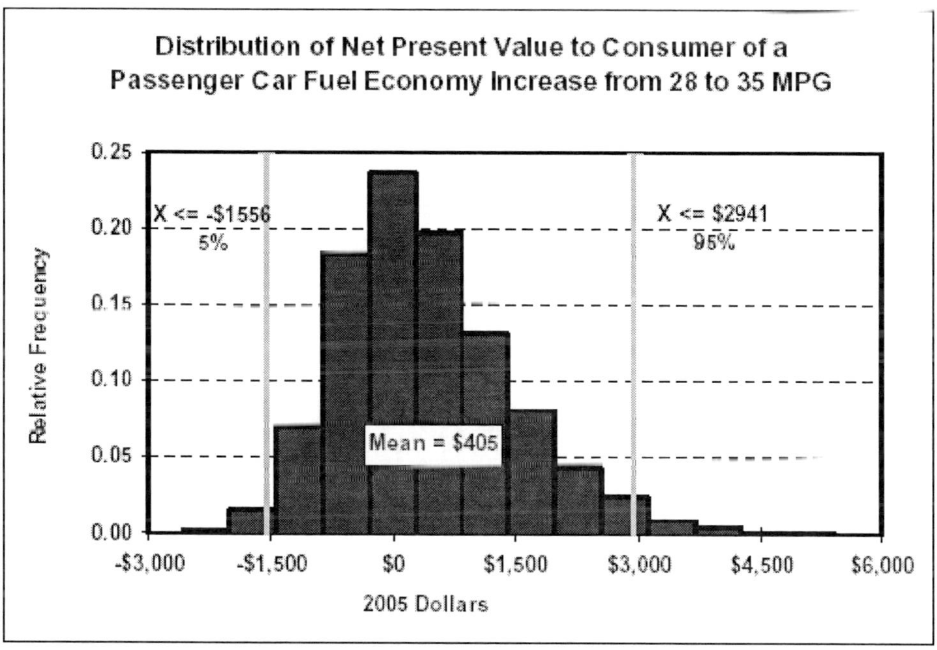

Figure 2. Distribution of Net Present Value to Consumer of an Increase in Passenger Car Fuel Economy from 28 to 35 MPG. (Greene, German and Delucchi, 2009)

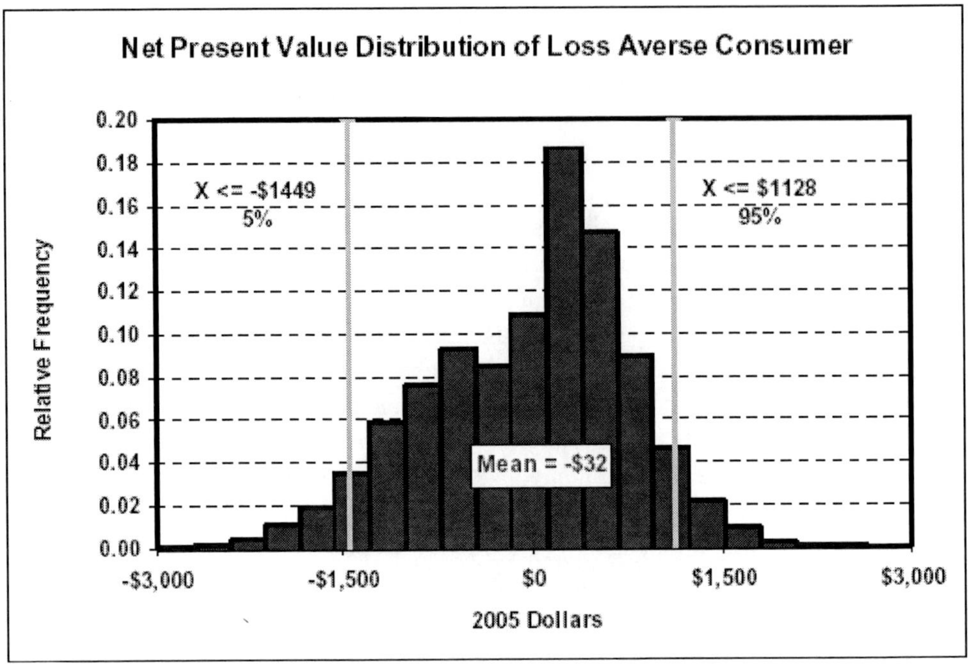

Figure 3. Perceived Value Distribution of the 25% Increase in Passenger Car Fuel Economy for a Loss-Averse Consumer. (Greene, German and Delucchi, 2009)

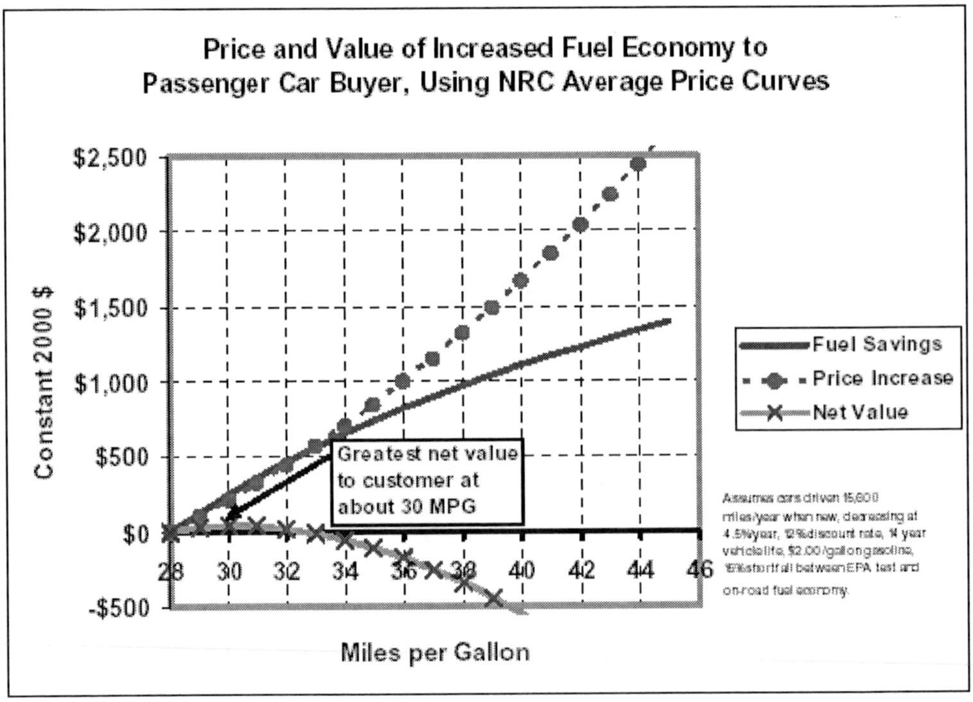

Figure 4. Private Cost and Expected Benefit of Increasing Passenger Car Fuel Economy Using a Simple 3-Year Payback Rule. (Greene, German and Delucchi, 2009)

The implications of uncertainty and loss-aversion match up almost exactly with the views expressed to the 2002 NRC committee by auto manufacturers, who stated that consumers were willing to pay only for technologies which paid back their cost in 2 to 4 years. If one assumes that consumers value future fuel savings using a simple 3-year payback rule, future fuel savings are undervalued by a factor of 2, or more. Again using the same cost data and assumptions of the 2002 NRC report, we find an expected value of fuel economy improvement of almost zero from 28 to 35 MPG (Figure 4). The undervaluing of energy efficiency (relative to expected savings) due to uncertainty and loss-aversion is very likely pervasive, affecting not only automobiles but all energy using consumer durable goods. It also almost certainly discourages appropriate levels of investment in energy efficiency research and development.

The phenomenon of uncertainty and loss-aversion in the market for fuel economy does not constitute a market failure in the usual sense. Rather it exacerbates the market failures of environmental externalities and oil market monopolization and its energy security consequences. It weakens the market response to fuel price signals, such as a tax on gasoline. At the same time it creates an opportunity for public policy to achieve a greater response than would be possible with externality pricing alone through the use of regulations or fiscal incentives to promote vehicle efficiency and GHG mitigation.

Examples of Successful Technology-Neutral Policies in Transportation

Motor vehicle emissions standards established by the State of California and the U.S. Environmental Protection Agency pursuant to the Clean Air Act, set performance goals for criteria pollutants but did not specify the technologies that should be used to achieve them. The motor vehicle industry responded with unanticipated technologies, such as the three-way catalyst, multi-point fuel injection and computerized control of combustion that reduced emissions by orders of magnitude. A passenger car meeting California's SULEV standard in 2005 emitted one one-thousandth (0.00 1) of the smog-producing hydrocarbons of a passenger car manufactured in 1960 (Sakai, 2009).

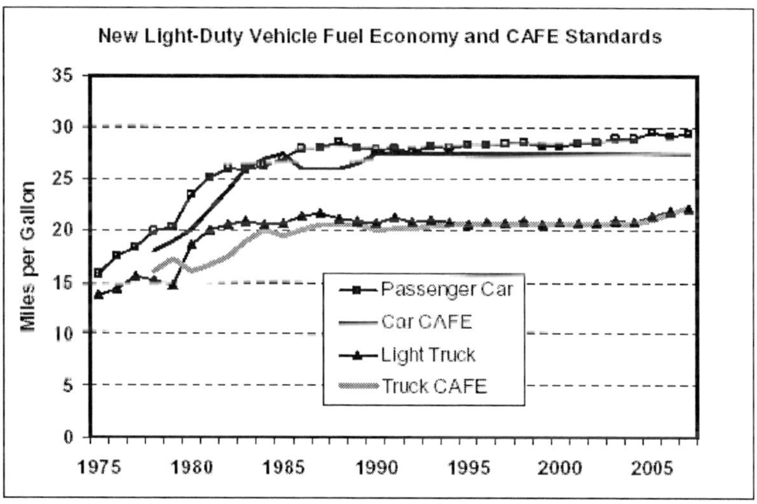

Source: U.S. EPA (2008) "Light-Duty Automotive Technology and Fuel Economy Trends: 1975-2008.

Figure 5. New Passenger Car and Light Truck Fuel Economy and Standards.

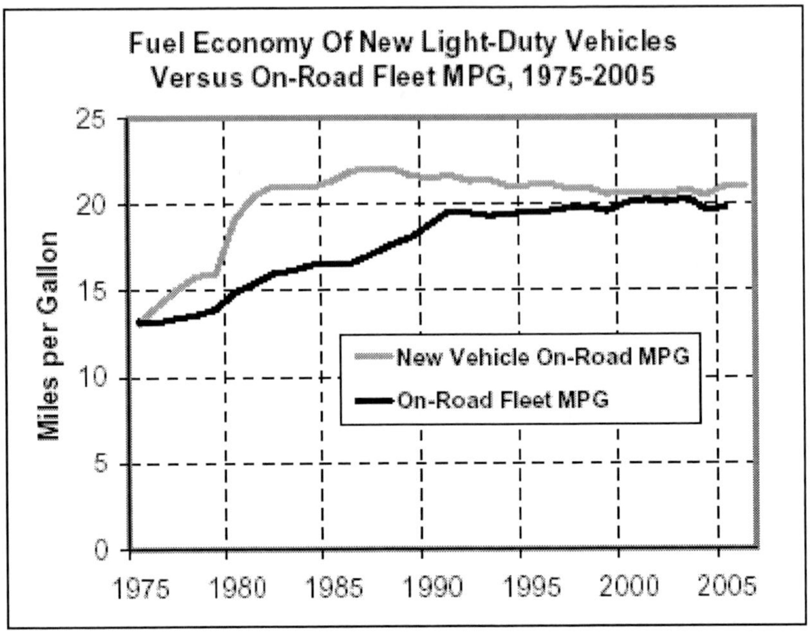

Sources: U.S. EPA (2008), table 1, and U.S. DOT/FHWA (2007), Highway Statistics 2007, table VM-1.

Figure 6. Fuel Economy of New Light-duty Vehicles & On-Road Fleet MPG, 1975-2007

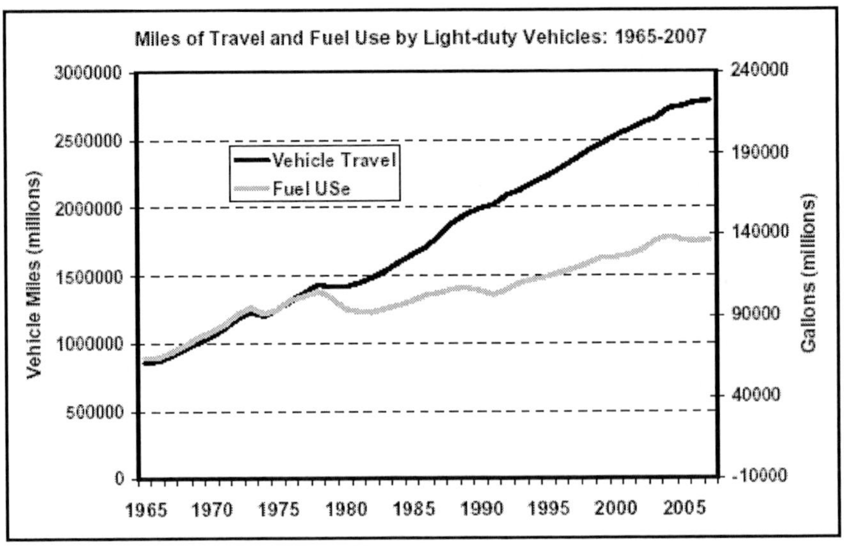

Source: U.S. DOT/FHWA (2007), *Highway Statistics 2007* and earlier, table VM-1.

Figure 7. Light-duty Vehicle Travel and Fuel Use: 1965-2007

The federal Corporate Average Fuel Economy standards likewise did not specify the technologies manufacturers should use to nearly double passenger car fuel economy over 1975 levels by 1985, and to increase light truck fuel economy by more than 50% (Figure 5).

Manufacturers responded with a range of technological solutions, from front wheel drive and lighter-weight unibody designs to reduced engine friction and 4- and 5-speed transmissions.

Improvements in new vehicle fuel economy gradually increased the fuel economy of the on-road fleet by more than 50%, as the vehicle stock turned over (Figure 6). The result was a clear decoupling of vehicle travel and energy use, beginning at about the same time the standards took effect in 1978 (Figure 7). U.S. motorists are today consuming on the order of 75 billion gallons less fuel each year than they would have had fuel use continued to increase in direct proportion to vehicle travel.

On the fuel side, California's clean fuel standards were initially pegged to the lower emission performance of methanol in comparison to gasoline. Fortunately, the standards were formulated in terms of pollutant emission requirements rather than picking methanol as the winning fuel. Energy companies responded by inventing reformulated gasolines that achieved the same environmental goals at lower cost.

Examples of technology-neutral fiscal incentives in the transport sector are scarcer.[2] Although it is a flawed policy, the U.S. gas-guzzler tax is nonetheless technology neutral, and appears to have had a powerful impact. The policy is flawed, in my view, because it applies only to passenger cars and not light trucks, and because it provides only for taxes on energy-intensive vehicles but no incentives for energy efficient vehicles. Still, it appears that the gas-guzzler tax convinced manufacturers to improve the fuel economy of larger passenger cars such that no mass-market passenger car has had to pay the tax (Greene et al., 2005).

New Options: Feebates and Carbon Prices

Can we replace our current patchwork of tax incentives for alternative fuels and vehicles (incentives for HEVs, PHEVs, natural gas, electric, flex-fuel, E85, for fuel economy, etc.) with a simpler yet more flexible and more efficient, technology-neutral incentive structure? The answer, I think, is a qualified yes. On the vehicle side, it should be possible to replace nearly all of our current incentives with a more effective unified vehicle incentive system, usually referred to as a "feebate" system. On the fuel side we can create a system for pricing carbon, whether it is a carbon tax or carbon cap-and-trade system. Both policies are technology neutral and can provide appropriate incentives for improving energy efficiency and de-carbonizing transportation's energy sources.

Feebates consist of a graduated rebate for energy efficient or low-GHG-emitting vehicles and a graduated levy on energy intensive or high-GHG-emitting vehicles, relative to a benchmark. The concept is very flexible: feebate systems can be formulated in an infinite number of ways. The simplest and perhaps most efficient, is to set a constant rate of rebate or tax per unit of petroleum use or GHG emissions per mile. A constant rate is economically efficient, in that it values every gallon of fuel saved or ton of GHGs mitigated equally.

The benchmark determines which vehicles pay a fee and which receive a rebate. It can be as simple as a single point for all vehicles or as complex as the reformed CAFE footprint function. Feebate systems can be designed to be revenue neutral, in which the fees finance the rebates, revenue enhancing or a net subsidy. At times when a stimulus to the automotive industry could benefit the entire economy, feebates could be structured as mainly or entirely a rebate system for energy efficient, low GHG emitting vehicles. By using an attribute-based benchmarking system, such as the NHTSA now uses in its reformed CAFE standards, equitable impacts on manufacturers could be designed into the system.

Regardless of where the benchmark is set, the feebate rate provides the market signal to automobile manufacturers to adopt advanced technology up to the point where the marginal cost equals the marginal value of fuel saved plus the marginal improvement in the feebate. The extra incentive of the feebate can be used to correct the market limitation described above. A feebate rate equivalent to an extra $1 per gallon of gasoline consumed over the life of a vehicle would have more leverage on new vehicle fuel economy than a $2 per gallon tax on gasoline.

Feebates differ from fuel economy standards in that as long as they are in effect they provide a continuing incentive to develop and apply new technology to improve fuel economy. Fuel economy standards must be periodically raised to stimulate continuous improvement. History indicates that this can be a significant problem. As Figure 5 shows, U.S. fuel economy standards were not significantly increased for more than two decades.

Unlike fuel economy standards, feebates also create an economic incentive for consumers to choose more efficient, lower GHG emitting vehicles. The strength and nature of the market signal, however, depends on precisely how the feebate benchmark is defined. A single benchmark for all light-duty vehicles will not only encourage consumers to select the more lower emitting vehicles within a size class but will also shift sales from larger to smaller vehicle classes. If the benchmark is defined as a footprint function, like the one used in the reformed CAFE system, there could be no incentive to choose smaller vehicles.

It is not yet clear, however, how best to use feebates as a complement or replacement for fuel economy or GHG standards. In theory, standards guarantee performance but not cost, while fiscal policies can assure cost but not performance. If we were omniscient, we could accomplish the same result with either policy. However, because there is very limited experience with feebate systems, it is not clear how much of the potential of technology to increase fuel economy or reduce GHG emissions would be traded-off by consumers and manufacturers for increased horsepower or size, or other energy-consuming features. These are trade-offs that fuel economy standards do not allow. More real-world experience with feebate systems will be accumulated in the next few years as the impacts of feebate-like systems implemented by France, the Netherlands, Spain and Sweden become known (Fulton, 2009). In addition, the Universities of California at Davis and Berkeley are conducting a comprehensive assessment of alternative feebate systems for California for the state's Air Resources Board. The study should provide useful insights about these and other practical issues.

The centerpiece of any climate policy should be establishing a meaningful way to price GHG emissions. It is well known that this can be done via a carbon tax or a carbon cap-and-trade system. There are pluses and minuses for either approach. I do not have a strong preference, however, if pressed I would give the edge to carbon cap-and-trade. Cap-and-trade will undoubtedly be more complex to administer but has the advantage that long term targets can be set, and long-term thinking is needed if we are to successfully cope with climate change. Transportation fuels should definitely be included in any carbon cap-and-trade system. Pricing carbon is not a replacement for efficiency standards but a useful complementary policy. A price on carbon will encourage energy companies to seek out ways to reduce the carbon content of the energy they supply to the transportation sector. It will also tend to increase the price of fuel, offsetting to a degree the small amount of increased driving that would otherwise be induced by increased vehicle efficiency (Small and Van Dender, 2007). Pricing carbon, however, is no panacea for transportation's energy problems.

\$50/tonCO_2 amounts to approximately \$0.50 per gallon of gasoline. Given the tendency of the market to undervalue future fuel savings, this is not nearly enough to stimulate the kinds of changes needed in our transportation system.[3]

How Large Should Incentives Be?

Incentives can be designed to reduce GHG emissions, reduce oil dependence, correct the fuel economy market limitation caused by uncertainty and loss aversion, or any combination of the three. To date, the subject of an optimal feebate rate has not been rigorously analyzed in this broad context. Rates established in past policies provide at least a few reference points. Feebate rates that follow from different carbon prices, oil security premiums and market corrections are then presented.

The U.S. gas-guzzler tax is specified in terms of dollar penalties per half MPG step below 22.5 MPG. Translated to gallons per mile, the average rate per 0.01 gallons per mile is approximately \$1,800 (Figure 8). In terms of dollars per gram of CO_2 per mile, this amounts to approximately \$20.

France's Bonus/Malus system corresponds to a rate of approximately \$1,500 per 0.01 gallons per mile, or approximately \$17 per gCO_2/mi (Figure 9). This rate does not apply to the final step for vehicles below 60gCO_2/km. The intention of the French government in setting a much higher incentive for the lowest emitting vehicles was to provide a strong incentive for developing advanced technologies such as PHEVs, EVs and FCVs. As noted above, such additional incentives are likely to be necessary during the early phase of a transition to a fundamentally different energy source for motor vehicles, such as hydrogen or electricity. Modifying a feebate rate curve to provide a greater incentive for advanced, near-zero-emission technologies is one approach to addressing the early barriers to a fundamentally different source of energy for transportation.

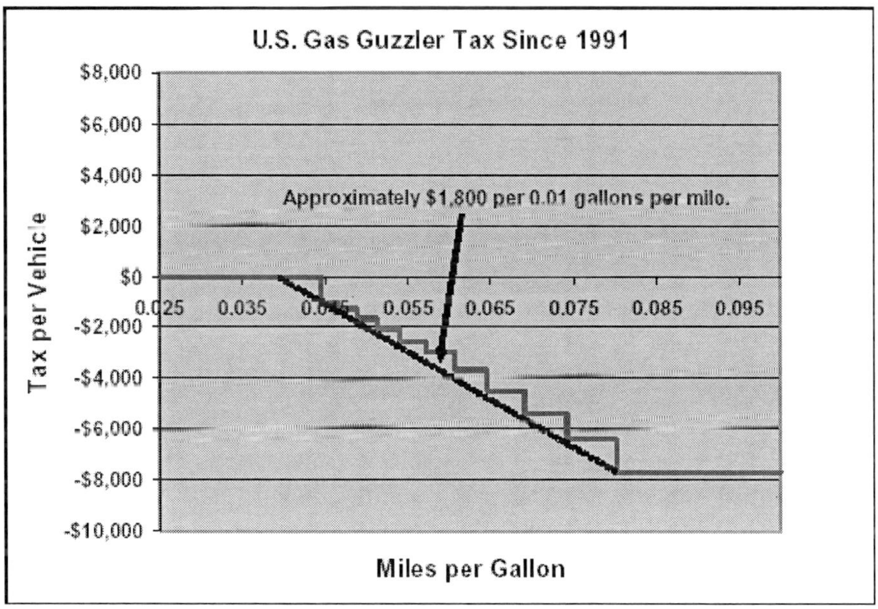

Figure 8. U.S. Gas-Guzzler Tax as an Fuel Efficiency Incentive Rate

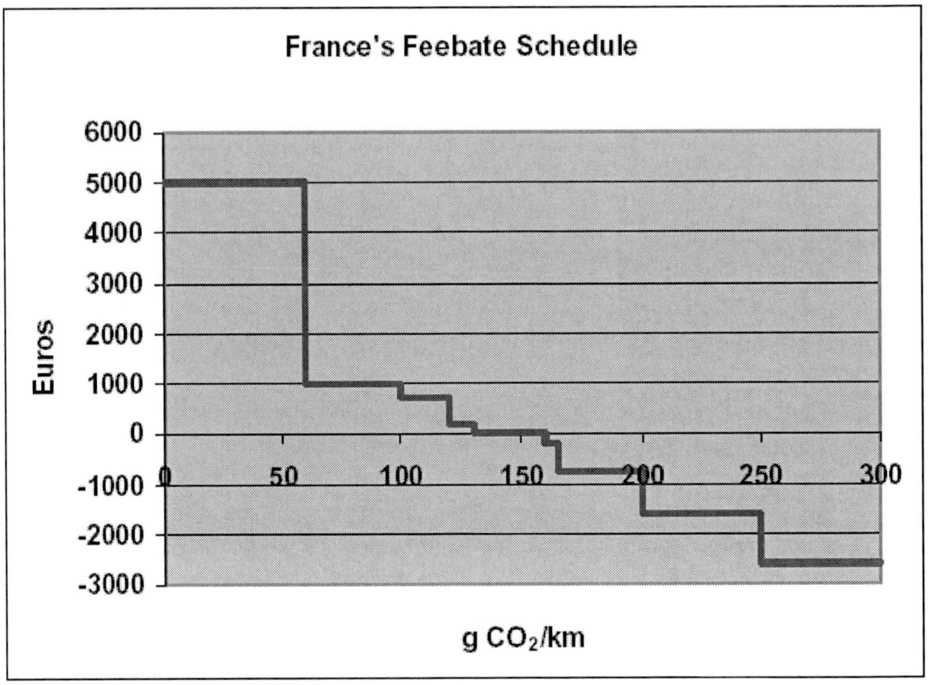

Figure 9. France's Bonus/Malus GHG Incentive System

While existing feebate-like systems provide useful reference points, it is more valuable for policy-making purposes to relate alternative feebate rates to equivalent carbon prices or oil consumption premiums. Table 1 shows a series of feebate rates from \$1,500 per 0.01 gallon per mile to \$2,500 per 0.01 gallon per mile and their equivalents in terms of carbon prices and per gallon surcharges on gasoline. Key assumptions used in the calculations are provided in a footnote to the table. A feebate rate of \$500 per 0.01 gallon per mile equates to a charge of \$0.47 per gallon of gasoline consumed over the life of the vehicle. If interpreted as a tax on CO_2 emissions, it equates to \$5.69 per gCO_2/mile or \$53 per metric ton of CO_2. Because the feebate system has 2.5 times the leverage on vehicle fuel economy as a gasoline tax, a gas tax of \$1.18 per gallon would be needed to equal the impact of the \$500 feebate rate.

The numbers in Table 1 assume that the full weight of the feebate is attributed to either an equivalent gasoline tax or a carbon price. More appropriately it should be shared between these objectives. Thus a \$2,000 feebate rate could be interpreted as a \$1/gal. gasoline tax and \$1 00/tCO_2 carbon price. However, neither calculation reflects the role of the feebate system in overcoming the problem of uncertainty and loss aversion. Given the undervaluing of future fuel savings by 2.5, a feebate equivalent to \$1.20/gal. would be needed just to overcome the uncertainty/loss-aversion effect. This may explain why the United States and France chose feebate rates in the vicinity of \$1 ,500-\$2,000 per 0.01 gallon per mile.

Table 1. Alternative Feebate Rates and Their Equivalencies in Terms of Externality Costs, Oil Consumption Premiums and Correcting the Uncertainty Loss-Aversion Problem

Feebate Rate \$/0.01gal/mi	Equivalent \$ per Lifetime PV Gal. \$/gal	Feebate Rate \$/gCO_2/mi	Equivalent Carbon Price \$/tCO_2	Gasoline Tax of Equal Impact \$/gallon
\$500	\$0.47	\$5.69	\$53	\$1.18
\$1,000	\$0.93	\$11.38	\$106	\$2.36
\$1,500	\$1.40	\$17.07	\$159	\$3.54
\$2,000	\$1.87	\$22.76	\$212	\$4.72
\$2,500	\$2.33	\$28.45	\$266	\$5.90

Assumes vehicle is driven 15,000 miles per year when new, declining at 4% per year, over a lifetime of 14 years. Future dollars are discounted at 7%/year.

Concluding Observations

The market's response to the problems of GHG emissions and oil dependence is not only hindered by the market failures of externalities and monopoly power but by the inherent uncertainty of future fuel savings and consumers' loss-averse behavior. As a consequence, fiscal incentives for increasing the energy efficiency of motor vehicles and reducing their GHG emissions can be especially effective policy tools. Technology neutral incentives have the dual advantages of allowing the greatest scope for innovation and harnessing market forces to select the most economically efficient solutions. As long as the incentives are in place they will provide a continuing incentive for firms to develop and implement, and for consumers to choose more energy efficient and lower emission vehicle technologies. Most, if not all of the existing incentives for energy efficient, low-emission vehicles technologies could be replaced by a consistent, economically efficient, technology neutral incentive system, such as feebates. What remains unclear at this point is how a comprehensive system of fiscal incentives should relate to a regulatory system targeting the same energy goals.

Feebate rates on the order of \$1,000 to \$2,000 per 0.01 gallons per mile and carbon prices in the vicinity of \$50 per ton of CO2 would very like be sufficient to stimulate research, development and implementation of advanced technologies and fuels that are not disruptive of the predominant petroleum fuel and internal combustion engine transportation system. Such economic incentives would probably not be adequate to initiate a sustainable transition to radically different energy sources for transportation, such as hydrogen or electricity. Hydrogen, in particular, will require a completely new energy supply infrastructure as well as entirely new propulsion systems for vehicles. Studies of what may be required for a transition to hydrogen vehicles (NRC, 2008; Greene et al., 2008) have concluded that even when the technological hurdles have been overcome, initiating a sustainable transition may require on the order of \$50 billion in subsidies to achieve learning-by-doing and economies of scale in vehicle production, and to provide sufficient fuel availability and diversity of vehicle choice to overcome the inertia of the petroleum-fueled, internal combustion engine system. During this early transition phase, which could easily last a decade, additional incentives especially for vehicles, are likely to be required. Such incentives could be provided as a special case or by modifying a feebate schedule to provide extra, temporary incentives for the lowest emission vehicles.

REFERENCES

Fulton, L. (2009). *Personal communication*, "Car CO_2 Taxation," International Energy Agency, Paris.

Greene, D. L. (2009). "Measuring Oil Security: Can the U.S. Achieve Oil Independence?" *Energy Policy*.

Greene, D.L., J. German and M.A. Delucchi. (2009). "Fuel Economy: The Case for Market Failure," in D. Sperling and J.S. Cannon, eds., *Reducing Climate Impacts in the Transportation Sector*, Springer Science+Business Media.

Greene, D. L., Leiby, P. N., James, B., Perez, J., Melendez, M., Milbrandt, A., Unnasch, S. & Hooks, M. (2008). *Transition to Hydrogen fuel Cell Vehicles & the Potential Hydrogen energy Infrastructure Requirements*, ORNL/TM-2008/30, Oak Ridge National Laboratory, Oak Ridge, Tennessee.

Greene, D. L., Patterson, P. D., Singh, M. & Li, J. (2005). "Feebates, Rebates and Gas-guzzler Taxes: A Study of Incentives for Increased Fuel Economy," *Energy Policy*, vol. *33*, pp. 757-775.

Leiby, P. N. (2007). *Estimating the Energy Security Benefits of Reduced U.S. Oil Imports*, ORNL/TM-2007/028, Oak Ridge National Laboratory, Oak Ridge, Tennessee, February.

National Research Council (NRC). (2008). *Transition to Alternative Transportation Technologies: A Focus on Hydrogen* (National Academies Press, Washington, DC).

National Research Council (NRC). (2002). *Effectiveness and Impacts of Corporate Average Fuel Economy Standards*, National Academies Press, Washington, DC.

Sakai, I. (2009). "Future of Transportation in the Carbon Constrained Environment: Technical and Political Perspectives of American Honda Motor," presentation at *The Johns Hopkins University School of Advanced International Studies*, March 25, Washington, DC.

Small, K. & Van Dender, K. (2007). "Fuel Efficiency and Motor Vehicle Travel: The Declining Rebound Effect," *The Energy Journal*, vol. *28*, no. 1, pp. 25-51.

Tversky, A. & Kahneman, D. (1992). "Advances in Prospect Theory: Cumulative Representation of Uncertainty," *Journal of Risk and Uncertainty*, vol. *5*, pp. 297-323.

(U.S. EPA) U.S. Environmental Protection Agency. (2008). *Light-Duty Automotive Technology and Fuel Economy Trends: 1975 through 2008*, EPA420-08-003, Office of Transportation and Air Quality, Ann Arbor, Michigan, September.

(U.S. DOT/FHWA) U.S. Department of Transportation, Federal Highway Administration. (2007). *Highway Statistics 2007*, FHWA-PL-08, Washington, DC.

End Notes

[1] Of course many other measures of merit are possible, as are combinations of measures.

[2] Certainly the federal excise taxes on motor fuels can be seen as a technology-neutral incentive to increase fuel economy, although their primary intent is to serve as a user fee to fund the highway system. By taxing the energy for transportation motor fuel taxes are effectively a tax on the amount of physical work done by transportation vehicles since, for constant energy efficiency, work done is directly related to energy use.

[3] This does not reflect the value of reducing oil dependence, which would justify a higher levy on petroleum fuels. However, as noted above, the nation's oil dependence problem is not an externality in the technical sense. Imposing a tax on oil, though helpful, is not a sufficient solution to the problem (Greene, 2009).

In: The Role of Taxes in Energy Production and Conservation ISBN: 978-1-61728-886-9
Editor: Bernadette M. Horton

Chapter 7

TESTIMONY OF JOHN M. URBANCHUK, DIRECTOR, LECG LLP, BEFORE THE SENATE FINANCE COMMITTEE, HEARING ON "TAX CODE'S ROLE IN ENERGY POLICY"*

Good morning, Chairman Baucus, Ranking Member Grassley, and Members of the Committee. My name is John M. Urbanchuk. I am a Director at LECG LLC, a global expert services consulting firm, where I specialize in agriculture and the economics of alternative fuels with particular emphasis on biofuels. I am pleased to be here this morning to discuss the role of the tax code in energy policy and whether technology-neutral energy incentives, including in the areas of fuels, vehicles, electricity, and efficiency, should be developed.

OPTIMAL STRUCTURE FOR ENERGY INCENTIVES

Experience, both in the U.S. and around the globe, has demonstrated that well crafted tax incentives are an effective means to encourage the production and use of renewable energy. Alternative sources of energy, specifically renewable forms, have been deemed in the national interest as a way of reducing dependence on imported energy and enhancing national security; improving environmental quality; and facilitating economic growth.

As with all tax policy, Congress should conduct prudent oversight to ensure that over time, various tax incentives continue to reflect the nation's tax and energy policy goals. As Congress considers both new tax incentives and revisits existing ones, there are ***three major factors*** that should be considered when determining the form and structure of an incentive.

* This is an edited, reformatted and augmented edition of a statement/testimony, given before the United States Senate Committee on Finance, dated April 23, 2009.

Industry Economics

The value of a tax incentive that is designed to encourage the production and use of a particular energy source should be set at responsible levels. Common sense dictates that the incentive should be structured in a manner that makes an activity economically viable but does not provide an unintended windfall for recipients. In addition, the value of the incentive should not create perverse incentives that encourage activities counter to responsible energy policy, or those that impede achievement of national energy policy goals.

Innovation and Technology Development

Energy tax policy should be structured in a manner that incentivizes and spurs the development of cleaner and more efficient ways to generate and distribute energy. A corollary to this is the role energy tax policy can play in promoting conservation and in helping direct the flow of private investment capital to cleaner and more efficient sources of energy, and industries with the potential for rapid commercialization. An additional consideration is the role of energy tax policy on stimulating job creation and economic growth, particularly in nascent industries such as wind and solar, and second generation biofuels. Thus, Congress should consider the potential to develop new technology, promote innovation and build needed infrastructure when considering energy tax policy.

Technology Neutrality

To the degree possible, energy tax incentives should be structured in a manner that treats competing technologies and processes in an equitable fashion. That said, Congress should set parameters, such as requiring various fuels to meet quality standards and specifications established by ASTM International, that ensure the desired policy goal of an incentive is being met.

These three equally important factors should be weighed when Congress considers energy tax policy. It is instructive to note that these three factors do not exist in isolation. Rather they interact with each other, thereby complicating the job of the policy maker.

A Technology-Neutral Tax Policy Would Counteract Existing Energy Policy

The issue of technology neutrality is particularly vexing, especially with regard to the development of alternative fuels. In its purest form a technology neutral energy tax policy would apply equally to all forms of energy and not give preferential treatment to one energy source over another. However, national policy as outlined by the Energy Independence and Security Act of 2007 (EPAC07), mandates the use of 36 billion gallons of renewable fuels by 2022 and provides important research and development incentives for solar and geothermal sources as well as for programs aimed at improving energy efficiency and conservation.

The development of energy tax policy that supports the national goals embodied by EPAC07 would effectively violate the basic premise of technology neutrality by providing favorable tax incentives for renewables such as cellulosic ethanol, advanced biofuel feedstocks including biomass biodiesel, solar, and geothermal. Specifically, EPAC07 establishes a cap of 15 billion gallons of ethanol from corn starch, calls for one billion gallons of biomass biodiesel, and requires the remaining 20 billion gallons of renewable fuels to come from cellulose and other advanced biofuel feedstocks. While we are on track to meet the 15 billion gallons of renewable fuels from traditional corn ethanol, the ability to produce the required amount of cellulose ethanol and biodiesel is, frankly, questionable.

The technology exists to process ethanol from cellulose feedstocks. However, commercialization of cellulosic ethanol remains a question of economics. While operating costs for cellulosic ethanol are expected to be lower than for corn ethanol, the capital investment necessary to build cellulosic ethanol facilities remain about five times that of grain-based facilities. The economic viability of any alternative fuel or energy source in today's environment of relatively low oil and gasoline prices and reduced demand as a result of recession is seriously threatened. This makes the continuance of existing tax incentives such as the Volumetric Ethanol Excise Tax Credit (VEETC), the Small Ethanol Producer Tax Credit and the biodiesel blenders excise tax credit all the more important in helping level the playing field for these alternatives to petroleum based fuels. The ethanol tax incentives, which continue through 2010, also play a major role in helping attract the investment capital needed to build and commercialize the second-generation (cellulose) ethanol industry.

The biodiesel tax incentive is scheduled to expire at the end of this year. In its absence the price of biodiesel will be significantly higher than petroleum diesel, further reducing demand and making it nearly impossible for biodiesel plants to produce fuel at a profit. Thus, it is safe to assume that if the biodiesel tax incentive lapses, biodiesel production in the U.S. will halt or at a minimum be severely curtailed, and the energy security, environmental, and job creation benefits that the nation realizes from biodiesel production will be lost.

Further, the short-term nature of the incentive under current law inadvertently sends the signal to the marketplace that the federal commitment to biodiesel is tenuous. At a time when market conditions are less than ideal and investor confidence is strained, the temporary nature of the incentive undermines overall confidence in the stability of the industry. A multi-year extension of a reformed tax incentive that is structured in a manner to promote a stable, viable domestic industry would address this situation and allow the U.S. to reap the multiple long-term benefits associated with the enhanced production and use of biodiesel.

Moving forward, in addition to the certainty provided by a multi-year extension, the biodiesel tax incentive can be reformed in a manner that will improve the form and function of the incentive. Specifically, changing the blenders excise tax credit to a production excise tax credit would improve administration of the credit for both taxpayers and the Treasury; help eliminate unintended abuses of the credit; and focus the incentive on the development of a domestic industry that is meeting the nation's energy needs.

Technology-Neutral Tax Policy Can Provide Unintended Consequences

Providing a technology-neutral tax policy can result in outcomes that provide an unintended windfall for recipients and create perverse incentives that encourages activities counter to responsible energy policy. Perhaps the most notable example of such an unintended consequence is the current outrage over "black liquor". The 2005 highway bill created a subsidy that provided a 50-cents-per-gallon tax credit for blending alternative fuels with traditional fossil fuels. In 2007 the law was later expanded to include other alternative fuels that would qualify for the credit as well as allowing "non-mobile" entities to qualify. This allowed the pulp and paper industry to claim the credit for blending a byproduct known as "black liquor," (already used as a fuel in plants) with a small amount of petroleum diesel. This has invoked the ire of environmental groups who claim that this use of the credit is actually encouraging the use of fossil fuels, and has increased the cost of the credit. The paper industry is only following longstanding industry practices and taking advantage of an existing technology-neutral tax incentive. Revoking the paper industry's eligibility for this incentive would violate the technology- neutrality concept.

Other examples of unintended consequences resulting from technology-neutral tax policies would include incentives to improve mileage or reduce emissions without regard to fuel type and incentives for the development of "clean coal" and coal-derived transportation fuels. The requirement to improve mileage without regard to fuel type could work to the disadvantage of flex-fuel vehicles that use renewable fuels, and end up increasing the use of petroleum based motor fuels compared to a policy that incented the use of renewable fuels. Incentives for coal could increase coal production and use, with the consequent environmental considerations.

These outcomes may be positive or negative, depending on the viewpoint of the interest group involved. The key point is that the full range of outcomes and consequences must be evaluated when tax policy is being developed.

Conclusion

Tax incentives have long supported public policies designed to stimulate the development of renewable energy markets and industries both in the U.S. and globally. Tax incentives are often complementary to other types of renewable energy incentive programs. They are powerful and highly flexible policy tools that can be targeted to encourage specific renewable energy technologies and to impact selected renewable energy market participants, especially when used in combination with other policy tools

The design of tax incentives deserves careful attention. The three equally important factors of industry economics, innovation and technology development, and technology-neutrality should be weighed when Congress considers energy tax policy. It is important that tax policy be consistent with and supportive of national energy policy goals and objectives and that careful thought be given to the full range of potential outcomes.

In: The Role of Taxes in Energy Production and Conservation ISBN: 978-1-61728-886-9
Editor: Bernadette M. Horton

Chapter 8

RENEWABLE ENERGY AND ENERGY EFFICIENCY TAX INCENTIVE RESOURCES*

Lynn J. Cunningham and Beth A. Roberts

SUMMARY

The following list of authoritative resources is designed to assist in responding to a broad range of constituent questions and concerns about renewable energy and energy efficiency tax incentives. Links are provided for the following: the full text of public laws establishing and extending federal renewable energy and energy efficiency incentives; federal, state, and local incentives resources; incentive resources grouped by technology type (solar, wind, geothermal, and biomass); CRS reports on this topic; and federal grants information resources. The last section of this chapter includes tables displaying popular incentives, the corresponding *U.S. Code* citations, and current expiration dates of those incentives.

This list reflects information that is currently available.

FULL TEXT OF TAX INCENTIVE LEGISLATION

Energy Policy Act of 2005 (EPACT) P.L. 109-58. The tax provisions are located in Title XIII.

Emergency Economic Stabilization Act of 2008 (EESA) P.L. 110-343. The tax provisions are located in Titles I, II, and III of Division B, the Energy Improvement and Extension Act of 2008.

* This is an edited, reformatted and augmented edition of a United States Congressional Research Service publication, Report R40455, dated March 23, 2009.

American Recovery and Reinvestment Act of 2009 (ARRA) P.L. 111-5. Conference Report with full text of the act (H.R. 1, as passed, P.L. 111-5) and the Joint Explanatory Statement. The tax provisions are located in Division B, Title I.

FEDERAL INCENTIVES

This section lists and describes several resources that contain information about federal incentives available to support energy efficiency and renewable energy.

Tax Incentives Assistance Project (TIAP)

http://www.energytaxincentives.org/

This website is sponsored by a number of government agencies, nonprofit groups, and other organizations. It focuses solely on information about federal tax incentives. Information is organized into categories for consumers, businesses, and builders/manufacturers. The site includes updates about enacted federal legislation and provides links to Internal Revenue Service (IRS) tax forms.

Environmental Protection Agency (EPA) Energy Star

http://www.energystar.gov/index.cfm?c=products.pr_tax_credits

This website has a page on "Federal Tax Credits for Energy Efficiency." The information on that page is organized into categories for consumers (home improvements, cars, solar energy, fuel cells), home builders, appliance manufacturers, and commercial buildings. The site includes a frequently asked questions (FAQ) section providing answers about energy efficiency tax credits.

Department of Energy (DOE) Financial Opportunities

http://www1.eere.energy.gov/financing/

This website is focused mainly on information about matching funds, grants, and financing. Information is organized into categories for consumers, business/industry/universities, inventors (small business), federal energy managers, states, and Native American tribes. The site includes a section on energy efficiency and consumer home financing.

U.S. Department of Energy Alternative Fuels and Advanced Vehicles Data Center (AFDC)

http://www.eere.energy.gov/afdc/

This website presents information about incentives for alternative fuels (renewable fuels and others) and vehicles. A key link provides access to "State and Federal Incentives and Laws." Incentives covered include grants, tax credits, loans, rebates, regulatory exemptions, fuel discounts, and technical assistance. Information on state incentives is made available through a national map and through summary tables organized by type of incentive, regulation, technology/fuel, and user. The information about state incentives is updated after each state legislature's session ends. Information about federal incentives is updated after pertinent legislation is enacted into law. Another link provides access to "Laws and Incentives Enactment History."

U.S. Department of Energy (DOE) Clean Cities Financial Opportunities

http://www1.eere.energy.gov/cleancities/financial_opps.html

This website presents information about incentives for alternative fuels and advanced technologies. A link to "Government Sources" provides information about funding opportunities through federal grant-making agencies (Grants.gov), Metropolitan Planning Organization (MPO), the Congestion Mitigation and Air Quality (CMAQ) Program, and various EPA programs. A link to "Solicitations" provides information about business funding opportunities that cover a variety of changing topics that have included plug-in hybrid vehicles, hydrogen vehicles, and transportation planning. Clean Cities Coordinators are available to help with funding applications.

Department of Housing and Urban Development (HUD) Energy Efficient Mortgages Program

http://www.hud.gov/offices/hsg/sfh/eem/energy-r.cfm

HUD's website provides information on Energy Efficient Mortgages. These mortgages can help homeowners finance the cost of adding energy-efficiency features to new or existing housing as part of their Federal Housing Authority-insured home purchase or refinance.

Alliance to Save Energy (ASE) Home and Vehicle Tax Credits

http://www.ase.org/content/article/detail/2654

ASE's website organizes information into categories on energy efficiency incentives for home improvements, hybrid vehicles, and solar energy. The site includes details on eligible equipment, credit limits, and credit expiration dates.

STATE AND LOCAL INCENTIVES

This section covers websites that list and describe state and local incentives available to support energy efficiency and renewable energy.

Database of State Incentives for Renewables and Efficiency (DSIRE) http://www.dsireusa.org/

DSIRE's website is sponsored by the Interstate Renewable Energy Council (IREC). IREC is a nonprofit organization focused on standards, guidelines, and other activities to support renewable energy. This site contains information about various types of energy efficiency and renewable energy financial incentives provided by state and local governments and utility companies. Summary data is accessed through a national map—and several additional special topic maps— that are linked to data on each state. Alternatively, the data can be searched by technology (solar, wind, geothermal), sector (residential, commercial/industrial, government, utility), and incentive type (tax credits, bonds, grants, loans), and eligible and implementing sectors. The site is updated weekly. Also, the homepage includes a map-link to a list of federal incentives.

U.S. Department of Energy (DOE) State Activities and Weatherization Assistance
http://www.eere.energy.gov/weatherization/

DOE's website on the Weatherization Assistance Program has information about how to apply for weatherization funding assistance. The site also has a "state activities" link, which provides information about state-level energy assistance programs.

INCENTIVES BY TECHNOLOGY TYPE

The following are links to resources by type of renewable energy.

Biomass

Database of State Incentives for Renewables and Energy Efficiency (DSIRE). Incentive for Biomass
http://www.dsireusa.org/

Energy Efficiency and Renewable Energy. Alternative Fuels and Advanced Vehicles Data Center. State and Federal Incentives and Laws
http://www.eere.energy.gov/afdc/progs/fed_summary.php/afdc/US/0

Environmental Protection Agency. Funding Database Biomass/Biogas
http://www.epa.gov/chp/funding

Geothermal

Database of State Incentives for Renewables and Energy Efficiency (DSIRE). Incentives for Geothermal Heat Pumps and Geothermal Electric
http://www.dsireusa.org/

Geothermal Heat Consortium – GeoExchange.org. New Federal Tax Credits for Geothermal Heat Pump Systems
http://www.geoexchange.org/component/content/article/90-new-federal-tax-credits-.html

Solar

Alliance to Save Energy. Energy-Efficiency Home and Vehicle Tax Credits. Solar Energy and Fuel Cell
http://www.ase.org/content/article/detail/2654#fuelcells_solar

Database of State Incentives for Renewables and Energy Efficiency (DSIRE). Incentives for Solar Technology http://www.dsireusa.org/

Energy Star. Federal Tax Incentives for Renewable Energy. Tax Incentives for Solar Energy Systems
http://www.energystar.gov/index.cfm?c=products.pr_tax_credits#s4

Solar Energy Industries Association. Solar Bills/Legislation
http://www.seia.org/cs/solar_bills

Solar Energy Industries Association. Frequently Asked Questions on the Solar Investment Tax Credit
http://www.seia.org/galleries/pdf/ITC_Frequently_Asked_Questions_10_9_08.pdf

Tax Incentives Assistance Project. Consumer Tax Incentives. Solar Energy Systems
http://www.energytaxincentives.org/consumers/

Tax Incentives Assistance Project. Businesses Tax Incentives. Solar Energy Systems
http://www.energytaxincentives.org/business/renewables.php

Wind

American Wind Energy Association. Legislative Affairs
http://www.awea.org/legislative/

Database of State Incentives for Renewables and Energy Efficiency (DSIRE). Incentives for Wind
http://www.dsireusa.org

CRS Reports on Federal Incentives

A number of CRS reports provide information about federal energy efficiency and/or renewable energy incentives:

Recent Legislation

CRS Report R40412, *Energy Provisions in the American Recovery and Reinvestment Act of 2009 (P.L. 111-5)*

General

CRS Report RL33831, *Energy Efficiency and Renewable Energy Legislation in the 110th Congress*

CRS Report RL33578, *Energy Tax Policy: History and Current Issues*

CRS Report RL34162, *Renewable Energy: Background and Issues for the 110th Congress*

Vehicles and Fuels

CRS Report RL32979, *Alcohol Fuels Tax Incentives*

CRS Report R40168, *Alternative Fuels and Advanced Technology Vehicles: Issues in Congress* CRS Report R40110, *Biofuels Incentives: A Summary of Federal Programs*

CRS Report RS22558, *Tax Credits for Hybrid Vehicles*

CRS Report RS22351, *Tax Incentives for Alternative Fuel and Advanced Technology Vehicles*

Wave, Tidal, In-Stream

CRS Report RL33883, *Issues Affecting Tidal, Wave, and In-Stream Generation Projects*

Wind Power

CRS Report RL34546, *Wind Power in the United States: Technology, Economic, and Policy Issues*

Popular Incentives Tables

Table 1. U.S. Code Citations and Expiration Dates for Popular Renewable Energy an Energy Efficiency Tax Incentives/Credits

Type of Credit	U.S. Code Citation	Expiration Date
Biofuels Production Tax Credit (PTC)	26 U.S.C. § 40A	12/31/2009
Business Solar and Fuel Cell Investment Tax Credit (ITC)	26 U.S.C. § 48(a)(3)	12/31/2016
Energy Efficiency Appliance Credit	26 U.S.C. § 45M	12/31/2008-12/31/2010 [a]
Energy Efficient Commercial Buildings Deduction	26 U.S.C. § 179D	12/31/2013
Energy Efficient New Homes Tax Credit for Homebuilders	26 U.S.C. § 45L	12/31/2009
Renewable Energy Production Tax Credit (PTC)	26 U.S.C. § 45	Wind (12/31/2012) Biomass, Geothermal,

		Hydroelectric, and Marine and Hydrokinetic (12/31/2013)
Residential Energy Efficient Property Credit (Solar	26 U.S.C. § 25D	12/31/2016

Source: *U.S. Code*

a. Depending on efficiency level of appliance

Table 2. Alternative Motor Vehicle Credit

Type of Credit	U.S. Code Citation	Expiration Date
Advanced Lean Burn Technology Motor Vehicle Credit	26 U.S.C. § 30B	12/31/2010
Fuel Cell Motor Vehicle Credit	26 U.S.C. § 30B	12/31/2014
New Qualified Plug-In Electrc Drive Motor Vehicles	26 U.S.C. § 30B	12/31/2014
Qualified Plug-In Electric Motor Vehicle Conversion Credit	26 U.S.C. § 30B	12/31/2011
Qualified Alternative Fuel Motor Vehicle Credit	26 U.S.C. § 30B	12/31/201
Qualified Hybrid Motor Vehicle Credit	26 U.S.C. § 30B	12/31/201

Source: U.S. Code

GRANT INFORMATION

Catalog of Federal Domestic Assistance (CFDA)

http://www.cfda.gov/

The CFDA is the primary source of federal grants program information, although actual funding depends upon annual congressional budget appropriations. The CFDA is available on the Internet. Many federal grants do not provide funding directly to individuals, but rather to states, local governments, universities, and tribal entities. Check the "applicant eligibility" section in the CFDA program description to see who may apply. Individuals may be eligible to apply for funds after they have been distributed at the state and local level, through their state energy offices or other contact listed in the CFDA program description.

Grants.gov

http://www.grants.gov

Federal grant funding opportunities are also posted on the website Grants.gov. Grants.gov enables grant seekers to electronically find and apply for competitive grants from all federal agencies.

CRS Reports on Grants

CRS Report RL34035, *Grants Work in a Congressional Office*

CRS Report RL32159, *How to Develop and Write a Grant Proposal*

CRS Report RL34012, *Resources for Grantseekers*

In: The Role of Taxes in Energy Production and Conservation ISBN: 978-1-61728-886-9
Editor: Bernadette M. Horton

Chapter 9

ENERGY TAX POLICY: HISTORY AND CURRENT ISSUES*

Salvatore Lazzari

SUMMARY

Historically, U.S. federal energy tax policy promoted the supply of oil and gas. However, the 1970s witnessed (1) a significant cutback in the oil and gas industry's tax preferences, (2) the imposition of new excise taxes on oil, and (3) the introduction of numerous tax preferences for energy conservation, the development of alternative fuels, and the commercialization of the technologies for producing these fuels (renewables such as solar, wind, and biomass, and nonconventional fossil fuels such as shale oil and coalbed methane). The Reagan Administration, using a free-market approach, advocated repeal of the windfall profit tax on oil and the repeal or phase-out of most energy tax preferences — for oil and gas, as well as alternative fuels. Due to the combined effects of the Economic Recovery Tax Act and the energy tax subsidies that had not been repealed, which together created negative effective tax rates in some cases, the actual energy tax policy differed from the stated policy. The George H. W. Bush and Bill Clinton years witnessed a return to a much more activist energy tax policy, with an emphasis on energy conservation and alternative fuels. While the original aim was to reduce demand for imported oil, energy tax policy was also increasingly viewed as a tool for achieving environmental and fiscal objectives. The Clinton Administration's energy tax policy emphasized the environmental benefits of reducing greenhouse gases and global climate change, but it will also be remembered for its failed proposal to enact a broadly based energy tax on Btus (British thermal units) and its 1993 across-the-board increase in motor fuels taxes of 4.3¢/gallon.

The 109th Congress enacted the Energy Policy Act of 2005 (P.L. 109-58), signed by President Bush on August 8, 2005, provided a net energy tax cut of $11.5 billion ($14.5 billion gross energy tax cuts, less $3 billion of energy tax increases) for fossil fuels and

---

* This is an edited, reformatted and augmented edition of a United States Congressional Research Service publication, Report RL33578, dated October 30, 2008.

electricity, as well as for energy efficiency, and for several types of alternative and renewable resources, such as solar and geothermal. The Tax Relief and Health Care Act of 2006 (P.L. 109-432), enacted in December 2006, provided for one-year extensions of these provisions. The current energy tax structure favors tax incentives for alternative and renewable fuels supply relative to energy from conventional fossil fuels, and this posture was accentuated under the Energy Policy Act of 2005.

On October 3, President Bush signed the Economic Stabilization Act of 2008 (P.L. 110-343), which includes $17 billion in energy tax incentives, primarily extensions of pre-existing provisions, but also including several new energy tax incentives: $10.9 billion in renewable energy tax incentives aimed at clean energy production, $2.6 billion in incentives targeted toward cleaner vehicles and fuels, and $3.5 billion in tax breaks to promote energy conservation and energy efficiency. The cost of the energy tax extenders legislation is fully financed, or paid for, by raising taxes on the oil and gas industry (mostly by reducing oil and gas tax breaks) and by other tax increases. The oil and gas tax increases comprise cutbacks in the IRC § 199 manufacturing deduction for income attributable to oil and gas production, which will be frozen at 6% (rather than increasing to 9% as scheduled), reforming the foreign tax credit provisions, and by increasing the per-barrel tax rate on refinery crude oil under the Oil Spill Liability Trust Fund provisions.

INTRODUCTION

Energy tax policy involves the use of the government's main fiscal instruments — taxes (financial disincentives) and tax subsidies (or incentives) — to alter the allocation or configuration of energy resources. In theory, energy taxes and subsidies, like tax policy instruments in general, are intended either to correct a problem or distortion in the energy markets or to achieve some social, economic (efficiency, equity, or even macroeconomic), environmental, or fiscal objective. In practice, however, energy tax policy in the United States is made in a political setting, being determined by the views and interests of the key players in this setting: politicians, special interest groups, bureaucrats, and academic scholars. This implies that the policy does not generally, if ever, adhere to the principles of economic or public finance theory alone; that more often than not, energy tax policy may compound existing distortions, rather than correct them.[1]

The idea of applying tax policy instruments to the energy markets is not new, but until the 1970s, energy tax policy had been little used, except for the oil and gas industry. Recurrent energy-related problems since the 1970s — oil embargoes, oil price and supply shocks, wide petroleum price variations and price spikes, large geographical price disparities, tight energy supplies, and rising oil import dependence, as well as increased concern for the environment — have caused policymakers to look toward energy taxes and subsidies with greater frequency.

Comprehensive energy policy legislation containing numerous tax incentives, and some tax increases on the oil industry, was signed on August 8, 2005 (P.L. 109- 58). The law, the Energy Policy Act of 2005, contained about $15 billion in energy tax incentives over 11 years, including numerous tax incentives for the supply of conventional fuels. However, record oil industry profits, due primarily to high crude oil and refined oil product prices, and the 2006 mid-term elections, which gave the control of the Congress to the Democratic Party, has changed the mood of policymakers. Instead of stimulating the traditional fuels industry — oil, gas, and electricity from coal — in addition to incentivizing alternative fuels and energy conservation, the mood now is to take away, or rescind, the 2005 tax incentives and use the money to further stimulate alternative fuels and energy conservation. A minor step in this direction was made, on May 17, 2006, when President Bush signed a $70 billion tax reconciliation bill (P.L. 109-222). This bill included a provision that further increased taxes on major integrated oil companies by extending the depreciation recovery period for geological and geophysical costs from two to five years (thus taking back some of the benefits enacted under the 2005 law). And currently, the major tax writing committees in both Houses are considering further, but more significant, tax increases on the oil and gas industry to fund additional tax cuts for the alternative fuels and energy conservation industries. These bills are being considered as part of the debate over new versions of comprehensive energy policy legislation in the 110th Congress (H.R. 6).

This chapter discusses the history, current posture, and outlook for federal energy tax policy. It also discusses current energy tax proposals and major energy tax provisions enacted in the 1 09th Congress. (For a general economic analysis of energy tax policy, see CRS Report RL3 0406, *Energy Tax Policy: An Economic Analysis*, by Salvatore Lazzari.)

BACKGROUND

The history of federal energy tax policy can be divided into four eras: the oil and gas period from 1916 to 1970, the energy crisis period of the 1 970s, the free- market era of the Reagan Administration, and the post-Reagan era — including the period since 1998, which has witnessed a plethora of energy tax proposals to address recurring energy market problems.

Energy Tax Policy from 1918 to 1970: Promoting Oil and Gas

Historically, federal energy tax policy was focused on increasing domestic oil and gas reserves and production; there were no tax incentives for energy conservation or for alternative fuels. Two oil/gas tax code preferences embodied this policy: (1) expensing of intangible drilling costs (IDCs) and dry hole costs, which was introduced in 1916, and (2) the percentage depletion allowance, first enacted in 1926 (coal was added in 1932).[2]

Expensing of IDCs (such as labor costs, material costs, supplies, and repairs associated with drilling a well) gave oil and gas producers the benefit of fully deducting from the first year's income ("writing off") a significant portion of the total costs of bringing a well into production, costs that would otherwise (i.e., in theory and under standard, accepted tax accounting methods) be capitalized (i.e., written off during the life of the well as income is earned). For dry holes, which comprised on average about 80% of all the wells drilled, the costs were also allowed to be deducted in the year drilled (expensed) and deducted against other types of income, which led to many tax shelters that benefitted primarily high-income taxpayers. Expensing accelerates tax deductions, defers tax liability, and encourages oil and gas prospecting, drilling, and the development of reserves.

The oil and gas percentage depletion allowance permitted oil and gas producers to claim 27.5% of revenue as a deduction for the cost of exhaustion or depletion of the deposit, allowing deductions in excess of capital investment (i.e, in excess of adjusted cost depletion) — the economically neutral method of capital recovery for the extractive industries. Percentage depletion encourages faster mineral development than cost depletion (the equivalent of depreciation of plants and equipment).

These and other tax subsidies discussed later (e.g., capital gains treatment of the sale of successful properties, the special exemption from the passive loss limitation rules, and special tax credits) reduced marginal effective tax rates in the oil and gas industries, reduced production costs, and increased investments in locating reserves (increased exploration). They also led to more profitable production and some acceleration of oil and gas production (increased rate of extraction), and more rapid depletion of energy resources than would otherwise occur. Such subsidies tend to channel resources into these activities that otherwise would be used for oil and gas activities abroad or for other economic activities in the United States. Relatively low oil prices encouraged petroleum consumption (as opposed to conservation) and inhibited the development of alternatives to fossil fuels, such as unconventional fuels and renewable forms of energy. Oil and gas production increased from 16% of total U.S. energy production in 1920 to 71.1% of total energy production in 1970 (the peak year).

Energy Tax Policy during the 1970s: Conservation and Alternative Fuels

Three developments during the 1970s caused a dramatic shift in the focus of federal energy tax policy. First, the large revenue losses associated with the oil and gas tax preferences became increasingly hard to justify in the face of increasing federal budget deficits — and in view of the longstanding economic arguments against the special tax treatment for oil and gas, as noted in the above paragraph. Second, heightened awareness of environmental pollution and concern for environmental degradation, and the increased importance of distributional issues in policy formulation (i.e., equity and fairness), lost the domestic oil and gas industry much political support. Thus, it became more difficult to justify percentage depletion and other subsidies, largely claimed by wealthy individuals and big vertically integrated oil companies. More importantly, during the 1 970s there were two energy crises: the oil embargo of 1973, also known as the first oil shock, and the Iranian Revolution in 1978-1979, which focused policymakers' attention on the problems (alleged "failures") in the energy markets and how these problems reverberated throughout the economy, causing stagflation, shortages, productivity problems, rising import dependence, and other economic and social problems.

These developments caused federal energy tax policy to shift from oil and gas supply toward energy conservation (reduced energy demand) and alternative energy sources.

Three broad actions were taken through the tax code to implement the new energy tax policy during the 1970s. First, the oil industry's two major tax preferences — expensing of IDCs and percentage depletion — were significantly reduced, particularly the percentage depletion allowance, which was eliminated for the major integrated oil companies and reduced for the remaining producers. Other oil and gas tax benefits were also cut back during this period. For example, oil- and gas-fired boilers used in steam generation (e.g., to generate electricity) could no longer qualify for accelerated depreciation as a result of the Energy Tax Act of 1978 (as discussed below).

The second broad policy action was the imposition of several new excise taxes penalizing the use of conventional fossil fuels, particularly oil and gas (and later coal). The Energy Tax Act of 1978 (ETA, P.L. 95-618) created a federal "gas guzzler" excise tax on the sale of automobiles with relatively low fuel economy ratings. This tax, which is still in effect, currently ranges from $1,000 for an automobile rated between 21.5 and 22.5 miles per gallon (mpg) to $7,700 for an automobile rated at less than 12.5 mpg. Chief among the taxes on oil was the windfall profit tax (WPT) enacted in 1980 (P.L. 96-223). The WPT imposed an excise tax of 15% to 70% on the difference between the market price of oil and a predetermined (adjusted) base price. This tax, which was repealed in 1988, was part of a political compromise that decontrolled oil prices. (Between 1971 and 1980, oil prices were controlled under President Nixon's Economic Stabilization Act of 1970 — the so-called "wage-price freeze.") (For more detail on the windfall profit tax on crude oil that was imposed from 1980 until its repeal in 1988, see CRS Report RL33305, *The Crude Oil Windfall Profit Tax of the 1980s: Implications for Current Energy Policy*, by Salvatore Lazzari.)

Another, but relatively small, excise tax on petroleum was instituted in 1980: the environmental excise tax on crude oil received at a U.S. refinery. This tax, part of the Comprehensive Environmental Response, Compensation, and Liability Act of 1980 (P.L. 96-5 10), otherwise known as the "Superfund" program, was designed to charge oil refineries for the cost of releasing any hazardous materials that resulted from the refining of crude oil. The tax rate was set initially at 0.79¢ ($0.0079) per barrel and was subsequently raised to 9.70¢ per barrel. This tax expired at the end of 1995, but legislation has been proposed since then to reinstate it as part of Superfund reauthorization.

The third broad action taken during the 1970s to implement the new and refocused energy tax policy was the introduction of numerous tax incentives or subsidies (e.g., special tax credits, deductions, exclusions) for energy conservation, the development of alternative fuels (renewable and nonconventional fuels), and the commercialization of energy efficiency and alternative fuels technologies. Most of these new tax subsidies were introduced as part of the Energy Tax Act of 1978 and expanded under the WPT, which also introduced additional new energy tax subsidies. The following list describes these:

- *Residential and Business Energy Tax Credits.* The ETA provided income tax credits for homeowners and businesses that invested in a variety of energy conservation products (e.g., insulation and other energy-conserving components) and for solar and wind energy equipment installed in a principal home or a business. The business energy tax credits were 10% to 15% of the investment in conservation or alternative fuels technologies, such as synthetic fuels, solar, wind, geothermal, and biomass. These tax credits were also expanded as part of the WPT, but they generally expired (except for business use of solar and geothermal technologies) as scheduled either in 1982 or 1985. A 15% investment tax credit for business use of solar and geothermal energy, which was made permanent, is all that remains of these tax credits.

- *Tax Subsidies for Alcohol Fuels.* The ETA also introduced the excise tax exemption for gasohol, recently at 5.2¢ per gallon out of a gasoline tax of 18.4¢/gal. Subsequent legislation extended the exemption and converted it into an immediate tax credit (currently at 51¢/gallon of *ethanol*).
- *Percentage Depletion for Geothermal.* The ETA made geothermal deposits eligible for the percentage depletion allowance, at the rate of 22%. Currently the rate is 15%.
- *§29 Tax Credit for Unconventional Fuels.* The 1980 WPT included a $3.00 (in 1979 dollars) production tax credit to stimulate the supply of selected unconventional fuels: oil from shale or tar sands, gas produced from geo-pressurized brine, Devonian shale, tight formations, or coalbed methane, gas from biomass, and synthetic fuels from coal. In current dollars this credit, which is still in effect for certain types of fuels, was $6.56 per barrel of liquid fuels and about $1.16 per thousand cubic feet (mcf) of gas in 2004.
- *Tax-Exempt Interest on Industrial Development Bonds.* The WPT made facilities for producing fuels from solid waste exempt from federal taxation of interest on industrial development bonds (IDBs). This exemption was for the benefit of the development of alcohol fuels produced from biomass, for solid-waste-to-energy facilities, for hydroelectric facilities, and for facilities for producing renewable energy. IDBs, which provide significant benefits to state and local electric utilities (public power), had become a popular source of financing for renewable energy projects.

Some of these incentives — for example, the residential energy tax credits — have since expired, but others remain and still new ones have been introduced, such as the §45 renewable electricity tax credit, which was introduced in 1992 and expanded under the American Jobs Creation Act of 2004 (P.L. 108-3 57). This approach toward energy tax policy — subsidizing a plethora of different forms of energy (both conventional and renewable) and providing incentives for diverse energy conservation (efficiency) technologies in as many sectors as possible — has been the paradigm followed by policymakers since the 1 970s. A significant increase in nontax interventions in the energy markets — laws and regulations, such as the Corporate Average Fuel Economy (CAFE) standards to reduce transportation fuel use, and other interventions through the budget and the credit markets — has also been a significant feature of energy policy since the 1970s. This included some of the most extensive energy legislation ever enacted.

Energy Tax Policy in the 1980s: The "Free-Market Approach"

The Reagan Administration opposed using the tax law to promote oil and gas development, energy conservation, or the supply of alternative fuels. The idea was to have a more neutral and less distortionary energy tax policy, which economic theory predicts would make energy markets work more efficiently and generate benefits to the general economy. The Reagan Administration believed that the responsibility for commercializing conservation and alternative energy technologies rested with the private sector and that high oil prices — real oil prices (corrected for inflation) were at historically high levels in 1981 and 1982 —

would be ample encouragement for the development of alternative energy resources. High oil prices in themselves create conservation incentives and stimulate oil and gas production.

President Reagan's free-market views were well known prior to his election. During the 1980 presidential campaign, he proposed repealing the WPT, deregulating oil and natural gas prices, and minimizing government intervention in the energy markets. The Reagan Administration's energy tax policy was professed more formally in several energy and tax policy studies, including its 1981 National Energy Policy Plan and the 1983 update to this plan; it culminated in a 1984 Treasury study on general tax reform, which also proposed fundamental reforms of federal energy tax policy. In terms of actual legislation, many of the Reagan Administration's objectives were realized, although as discussed below there were unintended effects.

In 1982, the business energy tax credits on most types of nonrenewable technologies — those enacted under the ETA of 1978 — were allowed to expire as scheduled; other business credits and the residential energy tax credits were allowed to expire at the end of 1985, also as scheduled. Only the tax credits for business solar, geothermal, ocean thermal, and biomass technologies were extended. As mentioned above, today the tax credit for business investment in solar and geothermal technologies, which has since been reduced to 10%, is all that remains of these tax credits. A final accomplishment was the repeal of the WPT, but not until 1988, the end of Reagan's second term. The Reagan Administration's other energy tax policy proposals, however, were not adopted. The tax incentives for oil and gas were not eliminated, although they were pared back as part of the Tax Reform Act (TRA) of 1986.

Although the Reagan Administration's objective was to create a free-market energy policy, significant liberalization of the depreciation system and reduction in marginal tax rates — both the result of the Economic Recovery Tax Act of 1981 (ERTA, P.L. 97-34) — combined with the regular investment tax credit and the business energy investment tax credits, resulted in negative effective tax rates for many investments, including alternative energy investments, such as solar and synthetic fuels. Also, the retention of percentage depletion and expensing of IDCs (even at the reduced rates) rendered oil and gas investments still favored relative to investments in general.

Energy Tax Policy After 1988

After the Reagan Administration, several major energy and non-energy laws were enacted that amended the energy tax laws in several ways, some major.

- *Revenue Provisions of the Omnibus Reconciliation Act of 1990.* President George H.W. Bush's first major tax law included numerous energy tax incentives: (1) for conservation (and deficit reduction), the law increased the gasoline tax by 5¢/gallon and doubled the gas-guzzler tax, (2) for oil and gas, the law introduced a 10% tax credit for enhanced oil recovery expenditures, liberalized some of the restrictions on the percentage depletion allowance, and reduced the impact of the alternative minimum tax on oil and gas investments; and (3) for alternative fuels, the law expanded the §29 tax credit for unconventional fuels and introduced the tax credit for small producers of ethanol used as a motor fuel.

- *Energy Policy Act of 1992 (P.L. 102-486).* This broad energy measure introduced the §45 tax credit, at 1.5¢ per kilowatt hour, for electricity generated from wind and "closed-loop" biomass systems. (Poultry litter was added later.) For new facilities, this tax credit expired at the end of 2001 and again in 2003 but has been retroactively extended by recent tax legislation (as discussed below). In addition, the 1992 law (1) added an income tax deduction for the costs, up to $2,000, of clean-fuel powered vehicles; (2) liberalized the alcohol fuels tax exemption; (3) expanded the §29 production tax credit for nonconventional energy resources; and (4) liberalized the tax breaks for oil and gas.
- *Omnibus Budget Reconciliation Act of 1993 (P.L. 103-66).* President Clinton proposed a differential Btu tax on fossil fuels (a broadly based general tax primarily on oil, gas, and coal based on the British thermal units of heat output), which was dropped in favor of a broadly applied 4.3¢/gallon increase in the excise taxes on motor fuels, with revenues allocated for deficit reduction rather than the various trust funds.
- *Taxpayer Relief Act of 1997 (P.L. 105-34).* This law included a variety of excise tax provisions for motor fuels, of which some involved tax reductions on alternative transportation fuels, and some involved increases, such as on kerosene, which on balance further tilted energy tax policy toward alternative fuels.
- *Tax Relief and Extension Act.* Enacted as Title V of the Ticket to Work and Work Incentives Improvement Act of 1999 (P.L. 106- 170), it extended and liberalized the 1.5 ¢/kWh renewable electricity production tax credit, and renewed the suspension of the net income limit on the percentage depletion allowance for marginal oil and gas wells.

As this list suggests, the post-Reagan energy tax policy returned more to the interventionist course established during the 1970s and primarily was directed at energy conservation and alternative fuels, mostly for the purpose of reducing oil import dependence and enhancing energy security. However, there is an environmental twist to energy tax policy during this period, particularly in the Clinton years. Fiscal concerns, which for most of that period created a perennial search for more revenues to reduce budget deficits, have also driven energy tax policy proposals during the post-Reagan era. This is underscored by proposals, which have not been enacted, to impose broad-based energy taxes such as the Btu tax or the carbon tax to mitigate greenhouse gas emissions.

Another interesting feature of the post-Reagan energy tax policy is that while the primary focus continues to be energy conservation and alternative fuels, no energy tax legislation has been enacted during this period that does not also include some, relatively minor, tax relief for the oil and gas industry, either in the form of new tax incentives or liberalization of existing tax breaks (or both).

ENERGY TAX INCENTIVES IN COMPREHENSIVE ENERGY LEGISLATION SINCE 1998

Several negative energy market developments since about 1998, characterized by some as an "energy crisis," have led to congressional action on comprehensive energy proposals, which included numerous energy tax incentives.

Brief History of Comprehensive Energy Policy Proposals

Although the primary rationale for comprehensive energy legislation has historically been spiking petroleum prices, and to a lesser extent spiking natural gas and electricity prices, the origin of bills introduced in the late 1990s was the very low crude oil prices of that period. Domestic crude oil prices reached a low of just over $10 per barrel in the winter of 1998-1999, among the lowest crude oil prices in history after correcting for inflation. From 1986 to 1999, oil prices averaged about $17 per barrel, fluctuating between $12 and $20 per barrel. These low oil prices hurt oil producers, benefitted oil refiners, and encouraged consumption. They also served as a disincentive to conservation and investment in energy efficiency technologies and discouraged production of alternative fuels and renewable technologies. To address the low oil prices, there were many tax bills in the first session of the 106th Congress (1999) focused on production tax credits for marginal or stripper wells, but they also included carryback provisions for net operating losses, and other fossil fuels supply provisions.

By summer 1999, crude oil prices rose to about $20 per barrel, and peaked at more than $30 per barrel by summer 2000, causing higher gasoline, diesel, and heating oil prices. To address the effects of rising crude oil prices, legislative proposals again focused on production tax credits and other supply incentives. The rationale was not tax relief for a depressed industry but tax incentives to increase output, reduce prices, and provide price relief to consumers.

In addition to higher petroleum prices there were forces — some of which were understood (factors such as environmental regulations and pipeline breaks) and others that are still are not so clearly understood — that caused the prices of refined petroleum products to spike. In response, there were proposals in 2000 to either temporarily reduce or eliminate the federal excise tax on gasoline, diesel, and other special motor fuels. The proposals aimed to help consumers (including truckers) cushion the financial effect of the price spikes. The Midwest gasoline price spike in summer 2000 kept interest in these excise tax moratoria alive and generated interest in proposals for a windfall profit tax on oil companies, which, by then, were earning substantial profits from high prices.

Despite numerous bills to address these issues, no major energy tax bill was enacted in the 106th Congress. However, some minor amendments to energy tax provisions were enacted as part of nonenergy tax bills. This includes Title V of the Ticket to Work and Work Incentives Improvement Act of 1999 (P.L. 106-170). Also, the 106th Congress did enact a package of $500 million in loan guarantees for small independent oil and gas producers (P.L. 106-51).

Energy Tax Action in the 107th Congress

In early 2001, the 107th Congress faced a combination of fluctuating oil prices, an electricity crisis in California, and spiking natural gas prices. The gas prices had increased steadily in 2000 and reached $9 per thousand cubic feet (mcf) at the outset of the 107th Congress. At one point, spot market prices reached about $30 per mcf, the energy equivalent of $175 per barrel of oil. The combination of energy problems had developed into an "energy crisis," which prompted congressional action on a comprehensive energy policy bill — the first since 1992 — that included a significant expansion of energy tax incentives and subsidies and other energy policy measures.

In 2002, the House and Senate approved two distinct versions of an omnibus energy bill, H.R. 4. While there were substantial differences in the nontax provisions of the bill, the energy tax measures also differed significantly. The House bill proposed larger energy tax cuts, with some energy tax increases. It would have reduced energy taxes by about $36.5 billion over 10 years, in contrast to the Senate bill, which cut about $18.3 billion over 10 years, including about $5.1 billion in tax credits over 10 years for two mandates: a renewable energy portfolio standard ($0.3 billion) and a renewable fuel standard ($4.8 billion). The House version emphasized conventional fuels supply, including capital investment incentives to stimulate production and distribution of oil, natural gas, and electricity. This focus assumed that recent energy problems were due mainly to supply and capacity shortages driven by economic growth and low energy prices. In comparison, the Senate bill would have provided a much smaller amount of tax incentives for fossil fuels and nuclear power and somewhat fewer incentives for energy efficiency, but provided more incentives for alternative and renewable fuels. The conference committee on H.R. 4 could not resolve differences, so the bills were dropped on November 13, 2002.

Energy Tax Action in the 108th Congress

On the House side, on April 3, 2003, the Ways and Means Committee (WMC) voted 24-12 for an energy tax incentives bill (H.R. 1531) that was incorporated into H.R. 6 and approved by the House on April 11, 2003, by a vote of 247-175. The House version of H.R. 6 provided about $17.1 billion of energy tax incentives and included $83 million of non-energy tax increases, or offsets. This bill was a substantially scaled-down version of the House energy tax bill, H.R. 2511 (107th Congress), which was incorporated into H.R. 4, the House energy bill of the 107th Congress that never became law. After returning from the August 2003 recess, a House and Senate conference committee negotiated differences among provisions in three energy policy bills: the House and Senate versions of H.R. 6, and a substitute to the Senate Finance Committee (SFC) bill — a modified (or amended) version of S. 1149 substituted for Senate H.R. 6 in conference as S.Amdt. 1424 and S.Amdt. 1431.

On November 14, 2003, House and Senate conferees reconciled the few remaining differences over the two conference versions of H.R. 6, which primarily centered on several energy tax issues — ethanol tax subsidies, the §29 unconventional fuels tax credit, tax incentives for nuclear power, and clean coal. On November 18, 2003, the House approved, by a fairly wide margin (246-180), the conference report containing about $23.5 billion of energy tax incentives. However, the proposed ethanol mandate would further reduce energy tax receipts — the 10- year revenue loss was projected to be around $26 billion. On November 24, Senate Republicans put aside attempts to enact H.R. 6. A number of uneasy alliances pieced together to bridge contentious divides over regional issues as varied as electricity, fuel additives (MTBE), and natural gas subsidies, failed to secure the necessary 60 votes to overcome a Democratic filibuster before Congress's adjournment for the holiday season. This represented the third attempt to pass comprehensive energy legislation, a top priority for many Republicans in Congress and for President Bush.

Senator Domenici introduced a smaller energy bill as S. 2095 on February 12, 2004. S. 2095 included a slightly modified version of the amended energy tax bill S. 1149; the tax provisions of S. 2095 were added to the export tax repeal bill S. 1637, on April 5, 2004. The Senate approved S. 1637, with the energy tax measures, on May 11. H.R. 4520, the House version of the export tax repeal legislation, did not contain energy tax measures; they were incorporated into H.R. 6.

Some energy tax incentives were enacted on October 4, 2004, as part of the Working Families Tax Relief Act of 2004 (P.L. 108-311), a $146 billion package of middle class and business tax breaks. This legislation, which was signed into law on October 4, 2004, retroactively extended four energy tax subsidies: the §45 renewable tax credit, suspension of the 100% net income limitation for the oil and gas percentage depletion allowance, the $4,000 tax credit for electric vehicles, and the deduction for clean fuel vehicles (which ranges from $2,000 to $50,000). The §45 tax credit and the suspension of the 100% net income limitation had each expired on January 1, 2004; they were retroactively extended through December 31, 2005. The electric vehicle credit and the clean-vehicle income tax deduction were being phased out gradually beginning on January 1, 2004. P.L. 108-311 arrested the phase-down — providing 100% of the tax breaks — through 2005, but resumed it beginning on January 1, 2006, when only 25% of the tax break was available. (For more information, see CRS Report RL3 2265, *Expired and Expiring Energy Tax Incentives*, by Salvatore Lazzari.)

The American Jobs Creation Act of 2004 (P.L. 108-357), commonly referred to as the "FSC-ETI" or "jobs" bill, was enacted on October 22, 2004. It included about $5 billion in energy tax incentives.

Energy Action in the 109th Congress

The 109th Congress enacted the Energy Policy Act of 2005 (P.L. 109-58), which included the most extensive amendments to U.S. energy tax laws since 1992, and the Tax Relief and Health Care Act of 2006, which extended the energy tax subsidies enacted under the 2005 Energy Policy Act (EPACT05).

The Energy Policy Act of 2005 (P.L. 109-58)

On June 28, 2005, the Senate approved by an 85-12 vote a broadly based energy bill (H.R. 6) with an 11-year, $18.6 billion package of energy tax breaks tilted toward renewable energy resources and conservation. Joint Committee on Taxation figures released on June 28 show that the bill included about $0.2 billion in non-energy tax cuts and more than $4.7 billion in revenue offsets, meaning the bill had a total tax cut of $18.8 billion over 11 years, offset by the $4.7 billion in tax increases. The House energy bill, which included energy tax incentives totaling about $8.1 billion over 11 years, and no tax increases, was approved in April. This bill was weighted almost entirely toward fossil fuels and electricity supply. On July 27, 2005, the conference committee on H.R. 6 reached agreement on $11.1 billion of energy tax incentives, including $3 billion in tax increases (both energy and non-energy). The distribution of the cuts by type of fuel for each of the three versions of H.R. 6 is shown in Table 1.

One way to briefly compare the two measures is to compare revenue losses from the energy tax incentives alone and the percentage distribution by type of incentive as a percent of the net energy tax cuts (i.e., the columns marked "%" divided by the dollar figures in row 11). The net revenue losses over an 11-year time frame from FY2005 to FY2015 were estimated by the Joint Committee on Taxation. The total revenue losses are reported in two ways. The absolute dollar value of tax cuts over 11 years and the percentage distribution of total revenue losses by type of incentive for each measure.

Table 1. Comparison of Energy Tax Provisions the House, Senate, and Enacted Versions of H.R. 6 (P.L. 109-58): 11-Year Estimated Revenue Loss by Type of Incentive (in millions of dollars; percentage of total revenue losses)

	House H.R. 6		Senate H.R. 6		P.L. 109-58	
	$	%	$	%	$	%
INCENTIVES FOR FOSSIL FUELS SUPPLY						
(1) Oil & Gas Production	-1,525	18.9%	-1,416	7.6%	-1,132	7.8%
(2) Oil & Gas Refining and Distribution	-1,663	20.6%	-1,399	7.5%	-1,501	10.4%
(3) Coal	-1,490	18.4%	-3,003	16.2%	-2,948	20.3%
(4) Subtotal	-4,678	57.8%	-5,818	31.3%	-5,581	38.6%
ELECTRICITY RESTRUCTURING PROVISIONS						
(5) Nuclear	-1,313	16.2%	-278	1.5%	-1,571	10.9%
(6) Other	-1,529	18.9%	-475	2.6%	-1,549	10.7%
(7) Subtotal	-2,842	35.1%	-753	4.1%	-3,120	21.6%
INCENTIVES FOR EFFICIENCY, RENEWABLES, AND ALTERNATIVE FUELS						
(8) Energy Efficiency	-570	7.0%	-3,987	2 1.4%	-1,260	8.7%
(9) Renewable Energy & Alternative Fuels	0	0%	-8,031	43.2%	-4,500	3 1.1%
(10) Subtotal	-570	7.0%	-12,018	64.6%	-5,760	39.8%
(11) Net Energy Tax Cuts	-8,010	100%	-18,589	100%	-14,461	100.0%
(12) Non Energy Tax Cuts[a]	0		-213		-92	
(13) Total Energy and Non-Energy Tax Cuts	0		-18,802		-14,553	
(14) Energy Tax Increases[b]	0		0		+2,857	
(15) Other Tax Increases			+ 4,705		171	
(16) NET TAX CUTS	-8,010		-14,055		-11,525	

Source: CRS estimates based on Joint Tax Committee reports.

a. The conference report includes a provision to expand R&D for all energy activities. This provision is listed as a nonenergy tax cut to simplify the table.

b. Energy tax increases comprise the oil spill liability tax and the Leaking Underground Storage Tank financing rate, both of which are imposed on oil refineries. If these taxes are subtracted from the tax subsidies (row 2), the oil and gas refinery and distribution sector suffered a net tax increase of $1,356 ($2,857-$ 1501); if the taxes are subtracted from all of the industry's tax subsidies (rows 1 and 2), the industry experienced a net tax increase of $224 million ($2,857- $2,633). Also, the Tax Increase Prevention and Reconciliation Bill of 2006 (P.L. 109-222), enacted on May 17, 2006, increased taxes on the oil industry by about $189 million.

Table 1 shows that the conference report provided about $1.3 billion for energy efficiency and conservation, including a deduction for energy-efficient commercial property, fuel cells, and micro-turbines, and $4.5 billion in renewables incentives, including a two-year extension of the tax code §45 credit, renewable energy bonds, and business credits for solar. A $2.6 billion package of oil and gas incentives included seven-year depreciation for natural gas gathering lines, a refinery expensing provision, and a small refiner definition for refiner depletion. A nearly $3 billion coal package provided for an 84-month amortization for pollution control facilities and treatment of §29 as a general business credit. More than $3 billion in electricity incentives leaned more toward the House version, including provisions providing 15-year depreciation for transmission property, nuclear decommissioning provisions, and a nuclear electricity production tax credit. It also provided for the five-year carryback of net operating losses of certain electric utility companies. A Senate-passed tax credit to encourage the recycling of a variety of items, including paper, glass, plastics, and electronic products, was dropped from the final version of the energy bill (H.R. 6). Instead, conferees included a provision requiring the Treasury and Energy departments to conduct a study on recycling. The House approved the conference report on July 28, 2005; the Senate on June 28, 2005, one month later on July 28, 2005, clearing it for the President's signature on August 8 (P.L. 109-58).

Four revenue offsets were retained in the conference report: reinstatement of the Oil Spill Liability Trust Fund; extension of the Leaking Underground Storage Tank (LUST) trust fund rate, which would be expanded to all fuels; modification of the § 197 amortization, and a small increase in the excise taxes on tires. The offsets total roughly $3 billion compared with nearly $5 billion in the Senate-approved H.R. 6. Because the oil spill liability tax and the Leaking Underground Storage Tank financing taxes are imposed on oil refineries, the oil and gas refinery and distribution sector (row 2 of Table 1) received a net tax increase of $1,769 ($2,857-$1,088).

The Tax Increase Prevention and Reconciliation Act (P.L. 109-222)

After expanding energy tax incentives in the EPACT05, the 109th Congress moved to rescind oil and gas incentives, and even to raise energy taxes on oil and gas, in response to the high energy prices and resulting record oil and gas industry profits. Many bills were introduced in the 1 09th Congress to pare back or repeal the oil and gas industry tax subsidies and other loopholes, both those enacted under EPACT05 as well as those that preexisted EPACT05. Many of the bills focused on the oil and gas exploration and development (E&D) subsidy — expensing of intangible drilling costs (IDCs). This subsidy, which has been in existence since the early days of the income tax, is available to integrated and independent oil and gas companies, both large and small alike.[3] It is an exploration and development incentive, which allows the immediate tax write-off of what economically are capital costs, that is, the costs of creating a capital asset (the oil and gas well).

Public and congressional outcry over high crude oil and product prices, and the oil and gas industry's record profits, did lead to a paring back of one of EPACT05's tax subsidies: two-year amortization, rather than capitalization, of geological and geophysical (G&G) costs, including those associated with abandoned wells (dry holes). Prior to the EPACT05, G&G costs for dry holes were expensed in the first year and for successful wells they were capitalized, which is consistent with economic theory and accounting principles. The Tax Increase Prevention and Reconciliation Act, (P.L. 109-222), signed into law May 2006, reduced the value of the subsidy by raising the amortization period from two years to five years, still faster than the capitalization treatment before the 2005 act, but slower than the treatment under that act. The higher amortization period applies only to the major integrated oil companies — independent (unintegrated) oil companies may continue to amortize all G&G costs over two years — and it applies to abandoned as well as successful properties. This change increased taxes on major integrated oil companies by an estimated $189 million over 10 years, effectively rescinding about 20% of the nearly $1.1 billion 11-year tax for oil and gas production under EPACT05.

The Tax Relief and Health Care Act of 2006 (P.L. 109-432)

At the end of 2006, the 109th Congress enacted a tax extenders package that included extension of numerous renewable energy and excise tax provisions. Many of the renewable energy provision in this bill had already been extended under the Energy Policy Act of 2005 and were not set to expire until the end of 2007 or later. The Tax Relief and Health Care Act of 2006 provided for one-year extensions of these provisions.

CURRENT POSTURE OF ENERGY TAX POLICY

The above background discussion of energy tax policy may be conveniently summarized in Table 2, which shows current energy tax provisions — both special (or targeted) energy tax subsidies and targeted energy taxes — and related revenue effects. A minus sign ("-") indicates revenue losses, which means that the provision is a tax subsidy or incentive, intended to increase the subsidized activity (energy conservation measures or the supply of some alternative and renewable fuel or technology); no minus sign means that the provision is a tax, which means that it should reduce supply of, or demand for, the taxed activity (either conventional fuel supply, energy demand, or the demand for energy-using technologies, such as cars).

Note that the table defines those special or targeted tax subsidies or incentives as those that are due to provisions in the tax law that apply only to that particular industry and not to others. Thus, for example, in the case of the oil and gas industry, the table excludes tax subsidies and incentives of current law that may apply generally to all businesses but that may also confer tax benefits to it. There are numerous such provisions in the tax code; a complete listing of them is beyond the scope of this chapter. However, the following example illustrates the point: The current system of depreciation allows the writeoff of equipment and structures somewhat faster than would be the case under both general accounting principles and economic theory; the Joint Committee on Taxation treats the excess of depreciation deductions over the alternative depreciation system as a tax subsidy (or "tax expenditure"). In FY2006, the JCT estimates that the aggregate revenue loss from this accelerated depreciation deduction (including the expensing under IRC § 179) is $6.7 billion. A certain, but unknown, fraction of this revenue loss or tax benefits accrues to the domestic oil and gas industry, but separate estimates are unavailable.

Table 2. Current Energy Tax Incentives and Taxes: Estimated Revenue Effects FY2007 (in millions of dollars)

Category	Provision	Major Limitations	Revenue Effects FY2007
CONVENTIONAL FOSSIL FUELS SUPPLY (bpd = barrels per day; < indicates less than)			
Targeted Tax Subsidies			
% depletion — oil, gas, and coal	15% of sales (higher for marginal wells); 10% for coal	only for independents, up to 1,000 or equiv. bpd	- 1,200
expensing of intangi-ble drilling costs (IDCs) and explorat-ion and development costs — oil/gas and other fuels	IDCs 100% deductible in first year	corporations expense only 70% of IDCs; remaining 30% are amortized over 5 years	- 1,100[a]
amortization of geol-ogical and geophysi-cal costs for oil and gas	costs amortized over 2 years for both dry holes and successful wells	major integrated oil compa-nies must amortize such costs (for both abandoned and successful properties) over 7 years	- 100
expensing of refinery investments	deduction of 50% of the cost of qualified refinery property, in the taxable year in which the refinery is placed in service	must increase the capacity of an existing refinery by 5%; remaining 50% is depr-eciated; must be placed in service before January 1, 2012	- 26
incentives for small refiners to comply with EPA sulfur regulations	$2.10 credit per barrel of low-sulfur diesel, plus expensing of 75% of capital costs	credit limited to 25% of capital costs; expensing phases out for refining capacity of 155,000-205,000 barrels per day.	- < 50
Tax Credits for Enhanced Oil Recovery Costs (EOR)	IRC §43 provides for a 15% income tax credit for the costs of recovering domestic oil by qualified "enhanced oil recovery" (EOR) methods, to ext-ract oil that is too viscous to be extracted by conve-ntional primary and secondary water-flooding techniques.	The EOR credit is non ref-undable, and is allowable provided that the average wellhead price of crude oil (using West Texas Interm-ediate as the reference), in the year before credit is cla-imed, is below the statutor-ily established threshold price of $28 (as adjusted for inflation since 1990), in the year the credit is claimed.	- 200

Table 2. (Continued)

Category	Provision	Major Limitations	Revenue Effects FY2007
		With average wellhead oil prices for 2005 (about $65) well above the reference price (about $38) the EOR credit was not available.	
Marginal Production Tax Credit	A $3 tax credit is provid-ed per barrel of oil ($0.50 per thousand cubic feet (mcf)) of gas from margi-nal wells, and for heavy oil.	The credit phases out as oil prices rise from $15 to $18 per barrel (and as gas prices rise from $1.67 to $2.00/ thousand cubic feet), adjus-ted for inflation. The credit is limited to 25 bpd or equ-ivalent amount of gas and to 1,095 barrels per year or equivalent. Credit may be carried back up to 5 years. At 2005 oil and gas prices, the marginal production tax credit was not available.	0
nuclear decommissioning	liberalizes tax deductible contributions to a fund in advance of actual decommissioning	in general, the IRS sets limits on the annual amou-nts made to a nuclear decommissioning fund	- 600
electric utilities	allows net-operating los-ses (NOLs) to be carried back 5 years, as compared with 2 years for all other industries	only 20% of the NOLs in 2003-2005 qualify	- < 50
disposition of electri-city transmission property to implement FERC policy	capital gain recognized evenly over 8 years	proceeds must be reinvested in other electricity gener-ating assets	- < 50
tax credit for advanced nuclear power facilities	1.8¢/kWh tax credit	limited to 6,000 megawatts of aggregate capacity; each taxpayer's credit also has a per kWh or power limita-tion and an aggregate limitation	- < 50
credit for clean-coal technologies	20% for integrated gasi-fication combined cycle (IGCC) systems; 15% for other advanced coal technologies	each system has maximum aggregate dollar limits	- 100
Targeted Taxes			
black-lung coal excise taxes and abandoned mineland reclamation (AML) fees	$1.25/ton for underground coal ($0.90 for surface coal)	coal tax not to exceed 4.4% of sales price (2.2% for the AML fee)	900
oil spill liability trust fund excise tax	$0.05/barrel tax on every barrel of crude oil refined	moneys are allocated into a fund for cleaning up oil spills	150

ALTERNATIVE, UNCONVENTIONAL, AND RENEWABLE FUELS			
Targeted Tax Subsidies			
§29, production tax credit	$6.40/bar. of oil or ($1.13/mcf of gas)	biogas, coal synfuels, coalbed methane, etc.	- 4,500
credits for fuel ethanol and biodiesel	$0.51 blender's credit plus $0.10/gal small producer credit	for biomass ethanol only (e.g., from corn)	- 3,000
tax credit for clean-fuel refueling property	$30,000 tax credit for alternative fuel equipment	per location, per taxpayer (replaces a deduction)	- < 50
§45 credit for renewable electricity	1.8¢/kWh. (0.9¢ in some cases; $4.375/ton of refined coal	wind, closed-loop biomass, poultry waste, solar, geothermal, etc.	- 1,100
alternative fuel motor vehicle (AFV) tax credits	$400-$40,000 credit for each fuel cell, hybrid, lean burn and other AFVs	tax credit is function of ve-hicle weight, fuel economy, and lifetime fuel savings	- 300
exclusion of interest on state and local bonds	interest income exempt from tax	for hydroelectric or biomass facilities used to produce electricity	- 100
credits for biodiesel	$0.50/gal. of recycled biodiesel; $1.00/gal. for virgin biodiesel	sold at retail or used in a trade or business, applies to oils from vegetables or animal fats	- 122
credit for business solar and geothermal technologies	10% investment tax credit for businesses	utilities excluded	- < 100
tax credit for renew-able energy bonds	credit equals the credit rate times by the bond's face amount	proceeds must be used for renewable electricity projects. national limit of $1.2 billion in bonds	- < 50
ENERGY CONSERVATION			
Targeted Subsidies			
mass transit subsidies	exclusion of $105/month		- 192
manufac-turer's credit for energy	max credit is $50 for dishwashers, $175 for	amount of credit depends on energy efficiency,	- 100
efficient appliances deduction for the cost of energy efficient property in comm-ercial buildings	refrigerators, and $200 for clothes washers tax deduction of cost of envelope components, heating cooling systems, and lighting	ene-rgy savings, and varies by year; total annual credit is also limited total deductions cannot exceed $1.80/sq.ft.	- < 50
credit for energy efficiency improvements to existing homes	10% tax credit ($500/ home) on up to $5,000 of costs; $50-$300 credit for other items	max credit on windows is $200	- 300
exclusion for utility conservation subsidies	subsidies not taxable as income	any energy conservation measure	- < 50
Targeted Taxes			
fuels taxes (FY2006)	18.4¢/gal. on gasoline	4.4¢-24.4¢ for other fuels	35,000
gas-guzzler tax (FY2006)	$1,000-$7,700/ vehicle wei-ghing 6,000 lbs. or less	trucks and SUVs are exempt	201

Source: Joint Tax Committee estimates and Internal Revenue Service data.

Notes: A negative sign indicates a tax subsidy or incentive; no negative sign indicates an energy tax. NA denotes not available.

a. The revenue loss estimate excludes the benefit of expensing costs of dry tracts and dry holes, which includes expensing some things that would otherwise be capitalized. This is a normal feature of the tax code but confers special benefits on an industry where the cost of finding producing wells includes spending money on a lot that turn out dry. This is probably more important than IDCs or percentage depletion.

ENERGY TAX POLICY IN THE 110TH CONGRESS

Continued high crude oil and petroleum product prices and oil and gas industry profits, and the political realignment of the Congress resulting from the 2006 Congressional elections continued the energy policy shift toward increased taxes on the oil and gas industry, and the emphasis on energy conservation and alternative and renewable fuels rather than conventional hydrocarbons.[4] In the 110th Congress, the shift became reflected in proposals to reduce oil and gas production incentives or subsidies, which were initially incorporated into, but ultimately dropped from comprehensive energy policy legislation. In the debate over these two comprehensive energy bills, raising taxes on the oil and gas industry, by either repealing tax incentives enacted under EPACT05, by introducing new taxes on the industry, or by other means was a key objective, motivated by the feeling that additional tax incentives were unnecessary — record crude oil and gasoline prices and industry profits provides sufficient (if not excessive) incentives.

In early December 2007, it appeared that the congressional conferees had reached agreement on another comprehensive energy bill, the Energy Independence and Security Act (H.R. 6), and particularly on the controversial energy tax provisions. The Democratic leadership in the 110th Congress proposed to eliminate or reduce tax subsidies for oil and gas and use the additional revenues to increase funding for their energy policy priorities: energy efficiency and alternative and renewable fuels (i.e., reducing fossil fuel demand) rather than an energy (oil and gas) supply increase. In addition, congressional leaders wanted to extend many of the energy efficiency and renewable fuels tax incentives that either had expired or were about to expire.

The compromise on the energy tax title in H.R. 6 proposed to raise taxes by about $21 billion to fund extensions and liberalization of existing energy tax incentives. However, the Senate stripped the controversial tax title from its version of the comprehensive energy bill (H.R. 6) and then passed the bill (86-8) on December 13, 2007, leading to the President's signing of the Energy Independence and Security Act of 2007 (P.L. 110-140), on December 19, 2007. The only tax-related provisions that survived were (1) an extension of the Federal Unemployment Tax Act surtax for one year, raising about $1.5 billion, (2) higher penalties for failure to file partnership returns, increasing revenues by $655 million, and (3) an extension of the amortization period for geological and geophysical expenditures from five to seven years, raising $103 million in revenues. The latter provision was the only tax increase on the oil and gas industry in the final bill. Those three provisions would offset the $2.1 billion in lost excise tax revenues going into the federal Highway Trust Fund as a result of the implementation of the revised Corporate Average Fuel Economy standards. The decision to strip the much larger $21 billion tax title stemmed from a White House veto threat and the Senate's inability to get the votes required to end debate on the bill earlier in the day. Senate Majority Leader Harry Reid's (D-Nev.) effort to invoke cloture fell short by one vote, in a 59-40 tally.

Since then, the Congress had tried several times to pass energy tax legislation, and thus avoid the impending expiration of several popular energy tax incentives, such as the "wind" energy tax credit under Internal Revenue Code (IRC) §45, which, since its enactment in 1992, had lapsed three times only to be reinstated.[5] Several energy tax bills have passed the House but not the Senate, where on several occasions, the failure to invoke cloture failed to bring up the legislation for consideration. Senate Republicans objected to the idea of raising taxes to offset extension of expiring energy tax provisions, which they consider to be an extension of current tax policy rather than new tax policy. In addition, Senate Republicans objected to raising taxes on the oil and gas industry, such as by repealing the (IRC) § 199 deduction, and by streamlining the foreign tax credit for oil companies.[6] The Bush administration repeatedly threatened to veto these types of energy tax bills, in part because of their proposed increased taxes on the oil and gas industry.

H.R. 5351

Frustrated with the lack of action on energy tax legislation over the last two years, House Democrats introduced and approved several such bills, such as H.R. 5351, which was approved by the House on February 27, 2008. House Speaker Pelosi and other Democrats sent President Bush a letter February 28, 2008, urging him to reconsider his opposition to the Democratic renewable energy plan, arguing that their energy tax plan would "correct an imbalance in the tax code."

H.R. 6049

As noted, several times the House had approved energy tax legislation, and several times in the Senate such legislation failed a cloture vote and thus could not be brought to the floor for debate. The latest was H.R. 6049, the House tax extenders bill, which was approved by the House on May 21, 2008, but failed three cloture votes in the Senate.[7]

H.R. 6899

In the House, energy tax provisions were part of H.R. 6899, House Democratic leadership's draft of broad-based energy policy legislation, the Comprehensive American Energy Security and Consumer Protection Act. Passed on September 16, 2008, the bill reverses the long-standing opposition of Democratic leaders to expanding oil and gas drilling offshore by allowing oil and gas exploration and production in areas of the outer continental shelf that are currently off limits, except for waters in the Gulf of Mexico off the Florida coast. Under the bill, states could allow such drilling between 50 and 100 miles offshore, while the federal government could permit drilling from 100 to 200 miles offshore.[8] Revenue from the new offshore leases would be used to assist the development of alternative energy, and would not be shared by the adjacent coastal states. The bill also repeals the current ban on leasing federal lands for oil shale production if states enact laws providing for such leases and production. H.R. 6899 also enacts a renewable portfolio standard, a mandate or requirement that power companies must generate 15% of their energy from renewable sources by 2020.

The energy tax provisions in H.R. 6899 (Title XIII, the Energy Tax Incentives Act of 2008) are largely the same as those in H.R. 5351.

Substitute Amendment of S. 3478

In the Senate, legislative efforts on energy tax incentives and energy tax extenders had centered around S. 3478, the Energy Independence and Investment Act of 2008, a $40 billion energy tax bill offered by Finance Committee Chairman Max Baucus and ranking Republican Charles Grassley. Senate Majority Leader Harry Reid said on September 12 that S. 3478 was "must-pass" legislation. Reid told reporters the energy tax package, which includes extensions of tax incentives for renewable energy, should be prioritized even ahead of the broader energy policy bills being considered, and the rest of the non-energy tax extenders package. Reid said he hopes to bring the bill to the floor during the week of September 15, but noted that the schedule depends on whether Senate Republicans will agree to move to the legislation.[9]

Although most of the tax incentives in the bill are extensions of existing policy and are not controversial, the legislation would need to be paid for through new sources of revenue. One proposed offset — which had been previously blocked by some Republicans — would have completely repealed the IRC § 199 manufacturing deduction for the five major oil and gas producers, raising $13.9 billion over 10 years. The bill also would have included a new 13% excise tax on oil and natural gas pumped from the Outer Continental Shelf, a proposal to eliminate the distinction between foreign oil-and-gas extraction income and foreign oil-related income, and an extension and increase in the oil spill tax through the end of 2017. In total, tax increases on the oil and gas industry would account for $31 billion of the $40 billion total cost of the legislation. The final major offset would have come from a requirement on securities brokers to report on the cost basis for transactions they handle to the Internal Revenue Service, a provision expected to raise about $8 billion in new revenues over 10 years.

The Economic Stabilization Act of 2008 (P.L. 110-343)

As noted above, some Republicans had, in the past, objected to the idea of raising taxes to offset extension of expiring energy tax provisions, which they consider to be an extension of current tax policy rather than new tax policy. In addition, some Senate Republicans have objected to raising taxes on the oil and gas industry, particularly by repealing the IRC § 199 deduction. The Bush Administration threatened also to veto any energy tax bill that would increase taxes on the oil and gas industry. At this writing, it appears that inclusion of the §199 deduction repeal as an offset might preclude the energy tax bill from coming to the Senate floor — some believe that it would fail another cloture vote — so this provision might not survive the process.[10]

Given continued Republican opposition (including a possible Presidential veto), and to avoid another legislative impasse — a failed cloture vote — Senators Baucus and Grassley released a scaled-down version of S. 3478.[11] This energy tax extenders package (i.e., the substitute of S. 3478) is a substitute amendment to the previously House-approved energy tax extenders bill H.R. 6049; is valued at nearly $17 billion, less than half the size of S. 3478; and is fully offset. The modified draft bill would also raise revenue by increasing the tax burden on the oil industry. Unlike the original version of S. 3478, however, which would have repealed the §199 for major integrated oil companies completely, the substitute bill would freeze the value of the manufacturing deduction for all oil companies at 6%, the current rate. This modification is estimated to raise $4.9 billion over 10 years, about 2/3 less than complete repeal. Because of smaller tax increases, the bill's remaining provisions — measures to increase tax subsidies for renewable fuels and for energy efficiency — had to be cut back. Thus, the scaled-down bill drops the nuclear electricity production tax credit provision, scales back the §45 renewable electricity tax credit, and generally shortens the extension periods.

The substitute amendment of S. 3478 (S.Amdt. 5633) was added to H.R. 6049, which also includes an AMT patch, disaster tax relief, and extensions of (non-energy) individual and business tax provisions. It was passed by the Senate on September 23, by a vote of 93-2. There was only one change to the original Baucus/Grassley substitute version: Language extending a 10¢ per-gallon credit for small producers of alcohol fuels was eliminated in the final bill.

Finally, H.R. 6049, including the energy tax amendments approved by the Senate, were added to the economic stabilization legislation (H.R. 1424) as Subdivision B, the Energy Improvement and Extension Act of 2008. On October 3, President Bush signed this legislation, the Economic Stabilization Act of 2008 (P.L. 110-343), which includes $17 billion in energy tax incentives, primarily extensions of pre-existing provisions, but also including several new energy tax incentives: $10.9 billion in renewable energy tax incentives aimed at clean energy production, $2.6 billion in incentives targeted toward cleaner vehicles and fuels, and $3.5 billion in tax breaks to promote energy conservation and energy efficiency. The cost of the energy tax extenders legislation is fully financed, or paid for, by raising taxes on the oil and gas industry (mostly by reducing oil and gas tax breaks) and by other tax increases. The oil and gas tax increases comprise cutbacks in the IRC §199 manufacturing deduction for income attributable to oil and gas production, which will be frozen at 6% (rather than increasing to 9% as scheduled), reforming the foreign tax credit provisions, and by increasing the per-barrel tax rate on refinery crude oil under the Oil Spill Liability Trust Fund provisions.

Windfall Profit Tax Legislation

Over the past ten years, surging crude oil and petroleum product prices have increased oil and gas industry revenues and generated record profits particularly for the top five major integrated companies (also known as the "super-majors"): Exxon- Mobil, Royal Dutch Shell, BP, Chevron, and Conoco/Phillips. These companies, which reported a predominate share of those profits, generated over $100 billion dollars in profits on nearly $1.5 trillion of revenues in 2007. From 2003 to 2007, revenues increased by 51%; net income (profits) increased by 85%. Oil output for the five majors over this time period declined by over 2%, from 9.85 to 9.63 million barrels per day. Since oil industry income has been largely price driven, with no increase in output, and with little new production resulting from increased oil industry investment, many believe that a portion of the increased income over this period represents a windfall and unearned gain, i.e., income not earned by any additional effort on the part of the firms, but due primarily to record crude oil prices, which are set in the world oil marketplace.

Numerous bills have been introduced in the Congress over this period to impose a windfall profit tax (WPT) on oil. Most of the bills were introduced in the 109th and 110th Congresses, after the enactment of the Energy Policy Act of 2005, which provided additional oil and gas industry tax incentives, on top of the industry's traditional tax subsidies. S. 3044, for instance, would roll back $17 billion in existing tax breaks over 10 years for the largest oil companies and impose a 25% windfall profit tax on major oil companies; revenues would be earmarked to expanding renewable energy development. In general, an excise-tax based WPT, like the one in effect from 1980-1988, would increase marginal oil production costs, reduce domestic oil supply, and raise petroleum imports, making the United States more dependent on foreign oil, undermining goals of energy independence and energy security. By contrast, the income-tax based WPT would be more economically neutral (less distortionary) in the short-run: sizeable revenues could be raised without reducing domestic oil supplies, which means oil imports would not tend to increase. Neither the excise-tax based or income-tax based WPT are expected to have significant price effects: neither tax would increase the price of crude oil, which means that refined petroleum product prices, such as pump prices, would likely not tend to increase.

In lieu of these two types of WPT, an administratively simple way of increasing the tax burden on the oil industry, and therefore recouping some of the excess or windfall profits, particularly from major integrated producers, would raise the corporate tax rate, by, for instance repealing or reducing the domestic manufacturing activities deduction under IRC § 199. This deduction is presently 6% of a firm's net income and is available generally to all domestic manufacturing businesses (service firms are excluded), including almost all oil firms. Repealing this deduction for the major integrated oil companies, and freezing it at 6% for the remaining qualifying oil companies is estimated by the Joint Committee on Taxation to generate about $10 billion over 10 years.

For an analysis of windfall profit legislation, see CRS Report RL34689, *Oil Industry Financial Performance and the Windfall Profit Tax*, by Salvatore Lazzari and Robert Pirog.

Energy Tax Provisions in the Farm Bill (P.L. 110-234)

It should also be mentioned that there are several, relatively small, energy tax provisions in the farm bill (H.R. 2419), which was just recently enacted (P.L. 110- 234). These provisions, all intended to promote alternative and renewable fuels from agricultural resources.

For Additional Reading

U.S. Congress, Senate Budget Committee, *Tax Expenditures: Compendium of Background Material on Individual Provision*, Committee Print, December 2006, 109th Cong., 2nd sess.

U.S. Congress, Joint Tax Committee, "*Description of the Tax Provisions in H.R. 2776, The Renewable Energy and Energy Conservation Tax Act of 2007*," June 19, 2007 (JCX-35-07).

U.S. Congress, Joint Tax Committee, "*Description of the Chairman's Modification to the Provisions of the Energy Advancement and Investment Act of 2007*," June 19, 2007 (JCX-33-07).

U.S. Congress, Joint Tax Committee, *Description And Technical Explanation of the Conference Agreement of H.R. 6, Title XIII, "Energy Tax Policy Tax Incentives Act of 2005*," July 27, 2005.

CRS Report RS21935, *The Black Lung Excise Tax on Coal*, by Salvatore Lazzari.

CRS Report RL33302, *Energy Policy Act of 2005: Summary and Analysis of Enacted Provisions*, by Mark Holt and Carol Glover.

CRS Report RL30406, *Energy Tax Policy: An Economic Analysis*, by Salvatore Lazzari.

CRS Report RS22344, *The Gulf Opportunity Zone Act of 2005*, by Erika Lunder.

CRS Report RL33763, *Oil and Gas Tax Subsidies: Current Status and Analysis*, by Salvatore Lazzari.

CRS Report RS22558, *Tax Credits for Hybrid Vehicles*, by Salvatore Lazzari.

CRS Report RS22322, *Taxes and Fiscal Year 2006 Reconciliation: A Brief Summary*, by David L. Brumbaugh.

CRS Report RL33305, *The Crude Oil Windfall Profits Tax of the 1980s:Implications for Current Energy Policy*, by Salvatore Lazzari.

CRS Report RL34669, *Side-by-Side Comparison of Energy Tax Bills in the House (H.R. 6049) and Senate (S. 3478),* by Salvatore Lazzari.

CRS Report RL34676, *Side-by-Side Comparison of the Energy Tax Provisions of H.R. 6899 and the Proposed Substitute of S. 3478*, by Salvatore Lazzari.

CRS Report RL34689, *Oil Industry Financial Performance and the Windfall ProfitsTax*, by Salvatore Lazzari and Robert Pirog.

End Notes

[1] The theory underlying these distortions, and the nature of the distortions, is discussed in detail in a companion report: CRS Report RL30406, *Energy Tax Policy: An Economic Analysis*, by Salvatore Lazzari.

[2] Tax preferences are special tax provisions — such as tax credits, exemptions, exclusions, deductions, deferrals, or favorable tax rates that reduce tax rates for the preferred economic activity and favored taxpayers. Such preferences, also known as tax expenditures or tax subsidies, generally deviate from a neutral tax system and from generally accepted economic and accounting principles unless they are targeted to the correction of preexisting market distortions.

[3] As is discussed later in the report, many of the other remaining tax subsidies are only available to independent oil and gas producers, which, however, may be very large.

[4] There is an important economic distinction between a subsidy and a tax benefit. As is discussed elsewhere in this chapter, firms receive a variety of tax benefits that are not necessarily targeted subsidies (or tax expenditures) because they are available generally.

[5] See. U.S. Library of Congress. Congressional Research Service. *Extension of Expiring Energy Tax Provisions.* CRS Report RL32265 by Salvatore Lazzari.

[6] Enacted in 2004 as an export tax incentive, this provision allows a deduction, as a business expense, for a specified percentage of the qualified production activity's income (or profit) subject to a limit of 50% of the wages paid that are allocable to the domestic production during the taxable year. The deduction was 3% of income for 2006, is currently 6%, and is scheduled to increase to 9% when fully phased in by 2010.

[7] Several times in the 110th Congress, the Senate has not taken action on energy tax legislation due to the failure to invoke cloture on the motion to proceed to the House energy tax extenders bills. The first was June 10, 2008,

when the motion failed by a vote of 50-44; the second was on June 17, when the motion failed by a vote of 52-44; the third was July 29, when the cloture motion failed by a vote of 53 to 43. In addition, on July 30 the Senate rejected by a vote of 51 to 43 a motion to invoke cloture on a motion to proceed to debate S. 3335, Senator Baucus' energy tax bill.

[8] The House Democratic leadership's energy proposal is centered around opening the Outer Continental Shelf to oil and gas development. The OCS areas — the Atlantic OCS, Gulf of Mexico (GOM) OCS, Pacific OCS, and Alaska OCS — are the offshore lands under the jurisdiction of the U.S. government. Federal law allows or confirms state boundaries and jurisdiction over the continental shelf areas up to 3 nautical miles from the coastline, except that (in the GOM) Texas and Florida offshore boundaries extend up to 9 nautical miles from the coastline. Exclusive federal jurisdiction over resources of the shelf applies from state boundaries out to 200 miles from the U.S. coastline For a more detailed definition of the OCS and various governmental jurisdictions see CRS Report RL33404, *Offshore Oil and Gas Development: Legal Framework*, by Adam Vann. For a comparison of different proposals see CRS Report RL34667, *Outer Continental Shelf Leasing: Side-by-Side Comparison of Five Legislative Proposals*, by Marc Humphries.

[9] Bureau of National Affairs. *Daily Tax Report.* "Reid Says 'Must Pass' Energy Legislation Should be Handled Before Tax Extenders." September 15, 2008. P. G-5.

[10] Bureau of National Affairs. *Daily Tax Report.* "Plan to Bring Tax Extenders to Floor Scraps Section 199 Deduction Repeal for Oil Firms." September 17, 2008. P. G-13.

[11] The legislative text and summary of the substitute of S. 3478 are in: Bureau of National Affairs. *Daily Tax Report*. September 18, 2008.

INDEX

A

B

C

D

E